BrightRED Study Guide

CfE HIGHER

ENGINEERING SCIENCE

Paul MacBeath

First published in 2019 by:
Bright Red Publishing Ltd
Mitchelston Business Centre
Mitchelston Drive
Kirkcaldy
KY1 3NB

Reprinted with corrections in 2021 and 2023.

Copyright © Bright Red Publishing Ltd 2019

Cover image © Caleb Rutherford

All rights reserved. No part of this publication may be reproduced, stored in a retrieval system, or transmitted in any form or by any means, electronic, mechanical, photocopying, recording or otherwise, without prior permission in writing from the publisher.

The rights of Paul MacBeath to be identified as the author of this work have been asserted by them in accordance with Sections 77 and 78 of the Copyright, Designs and Patents Act 1988.

A CIP record for this book is available from the British Library.

ISBN 978-1-84948-316-2

With thanks to:
PDQ Digital Media Solutions Ltd, Bungay (layout), Ivor Normand (copy-edit).
Cover design and series book design by Caleb Rutherford – e i d e t i c.

Acknowledgements
Every effort has been made to seek all copyright-holders. If any have been overlooked, then Bright Red Publishing will be delighted to make the necessary arrangements.

Permission has been sought from all relevant copyright holders and Bright Red Publishing are grateful for the use of the following:

Lynette and Malcolm Johnson/Creative Commons (CC BY-SA 2.0)[1] (p 11); oomlout/Creative Commons (CC BY-SA 2.0)[1] (p 28); Dcaldero8983/Creative Commons (CC BY-SA 3.0)[2] (p 50); Barry Skeates/Creative Commons (CC BY 2.0)[3] (p 56); Paul Harrop/Creative Commons (CC BY-SA 2.0)[1] (p 59); Sebastian Ballard/Creative Commons (CC BY-SA 2.0)[1] (p 75); Yummifruitbat/Creative Commons (CC BY-SA 2.5)[4] (p 78); Ahodges7/Creative Commons (CC BY-SA 3.0)[2] (p 78); Skatesandy/Creative Commons (CC BY-SA 4.0)[5] (p 85).

[1] (CC BY-SA 2.0)   http://creativecommons.org/licenses/by-sa/2.0/
[2] (CC BY-SA 3.0)   http://creativecommons.org/licenses/by-sa/3.0/
[3] (CC BY 2.0)   http://creativecommons.org/licenses/by/2.0/
[4] (CC BY-SA 2.5)   http://creativecommons.org/licenses/by-sa/2.5/
[5] (CC BY-SA 4.0)   http://creativecommons.org/licenses/by-sa/4.0/

Printed and bound in the UK.

CONTENTS

INTRODUCING HIGHER ENGINEERING SCIENCE
Course Content and Assessment 4

ENGINEERING CONTEXTS AND CHALLENGES
Engineering.. 6
Sub-system Diagrams 8
Analogue Control Systems 10
Energy and Efficiency 12

ELECTRONICS AND CONTROL
Analogue Electronics: Electronic Circuits 14
Analogue Electronics: Voltage-Divider Circuits....... 16
Analogue Electronics: Operational Amplifiers 1 18
Analogue Electronics: Operational Amplifiers 2 20
Analogue Electronics: Operational Amplifiers 3 22
Analogue Electronics: Operational Amplifiers 4 24
Analogue Electronics: Transistors 26
Analogue Electronics: MOSFETs........................ 28
Digital Electronics: Logic Gates and Boolean Expressions... 30
Digital Electronics: Combinational Boolean 32
Digital Electronics: Combinational Logic Tables 34
Digital Electronics: Creating Boolean from Truth Tables 36
Digital Electronics: NAND Equivalents 38
Programmable Control: Flowcharting.................. 40
Programmable Control: Writing Code with PBASIC 1 42
Programmable Control: Writing Code with PBASIC 2 44
Programmable Control: Writing Code with Arduino 1 46
Programmable Control: Writing Code with Arduino 2 48
Programmable Control: Motor Control 50

MECHANISMS AND STRUCTURES
Mechanical Systems: Drive Systems 52
Mechanical Systems: Coupling Methods and Clutches... 54
Mechanical Systems: Friction.......................... 56
Mechanical Systems: Torque and Mechanical Power 58
Pneumatic Systems: Valves and Cylinders............. 60
Pneumatic Systems: Speed Control and Time Delays 62
Pneumatic Systems: Sequential Control............... 64
Pneumatic Systems: Automatic Circuits and Cascade Systems... 66
Structures: Moments and Reactions 68
Structures: Forces at Angles........................... 70
Structures: Resultants and Concurrent Forces......... 72
Structures: Complex Resolution of a Force 74
Structures: Hinged Supports 76
Structures: Framed Structures......................... 78
Structures: Nodal Analysis 1............................ 80
Structures: Nodal Analysis 2............................ 82
Materials: Material Properties 84
Materials: Stress and Strain............................ 86
Materials: Young's Modulus 88
Materials: Factor of Safety............................. 90

COURSE ASSESSMENT
Overview 92

APPENDICES
Glossary... 94
Index.. 95

INTRODUCING HIGHER ENGINEERING SCIENCE

COURSE CONTENT AND ASSESSMENT

SYLLABUS

The main purpose of this book is to help you improve your chances of success within the Higher Engineering Science course, and to act as a supplement to your learning in school. Regular studying of this book, and using it to strengthen your learning and understanding of what you are taught in class, will go a long way towards your success in the course.

The course consists of three main units, and each chapter within this book covers the content you will need to know and understand. These units are:

- Engineering Contexts and Challenges
- Electronics and Control
- Mechanisms and Structures.

Engineering Contexts and Challenges

This part of the course gives a broad context of engineering. As you complete this, you will develop a deep understanding of engineering concepts by exploring and analysing a range of complex engineered objects, engineering problems and their solutions. You will investigate existing and emerging technologies and engineering challenges, and consider any implications that will arise from the solutions.

Electronics and Control

Within this part of the course, you will study several key concepts and devices used within electronic control systems. These will include analogue, digital and programmable electronics. Your skills in problem-solving and evaluating will be developed further, through simulation and/or practical projects, and you will explore several different engineering problems and solutions in a range of different contexts.

Mechanisms and Structures

This unit is designed to help develop a deep understanding of mechanisms and structures. Areas of this section may also be done through simulation and/or practical projects. Your problem-solving and evaluation skills will be strengthened through several different practical, simulated and investigative tasks.

ONLINE

Head to www.brightredbooks.net for a link to SQA documentation on the course.

COURSE ASSESSMENT

To achieve success in this course, you must be able to show that you can apply the knowledge and skills developed through the course, in both practical and theoretical contexts. To do this, your assessment will be broken into two separate parts:

Part 1: assignment (50 marks, 31% of the overall award)

Part 2: question paper (110 marks, 69% of the overall award).

The marks of these two components will be combined, and your award will be based on your total score. This means your total course assessment will be based on 160 marks.

Assignment

The assignment is designed so that you can demonstrate aspects of challenge and application in a practical context. You will apply knowledge and skills you have gained

DON'T FORGET

Make sure you read through the task and marking instructions beforehand to ensure you fully understand what is being asked of you, and you know how to succeed.

contd

Introducing Higher Engineering Science: Course Content and Assessment

from completing this course to solve a challenging practical engineering problem. This has 50 marks, and is 31% of your overall grade, so it is worthwhile putting in the effort to ensure it is completed to the best of your ability. Like the National 5 assignment, this will be completed under closed-book conditions and should take 8 hours to complete.

The assessment of this project will be similar to what you experienced in National 5. There will be one overarching project, but it will be broken into different tasks. Within this assignment, you may be asked to complete tasks focusing around five key areas:

Analysing the problem
This could be writing or reading a specification, drawing system diagrams and/or sub-system diagrams, or writing a description of a system.

Designing a possible solution
This may include reading or creating a flowchart or design mechanisms and structures, with calculations to prove that the solution works to given criteria.

Constructing/simulating a solution
This may include producing code and modelling/simulating solutions to problems in any electronic, mechanical or structural systems.

Testing the solution
This will include creating or completing test plans and then testing the potential solutions you have created.

Reporting and evaluation of the solution
This will include reporting on what you have done, and a reflection of your development process. It will also include an evaluation of any solution against a set specification.

Question Paper

The purpose of the exam is to assess your breadth and application of knowledge from across all the units. Here you will display your depth of understanding, and it will give you the opportunity to apply this knowledge and understanding to answer a selection of challenging questions.

The question paper will have 110 marks, which is 69% of the total mark, and is designed to give you the opportunity to demonstrate your skills, knowledge and understanding in the areas in this table:

The question paper will have two different sections.

Section 1 will be out of 20 marks, and will consist of short-answer questions.

Section 2 will be out of 90 marks, and will be made up from longer, more challenging questions that may combine several areas of the course.

Area	Marks
Engineering roles and disciplines, the impacts of engineering, the systems approach, energy and efficiency	10–17
Analogue electronic control systems	20–35
Digital electronic control systems	15–25
Drive systems and pneumatics	10–20
Structures and forces	15–25
Materials	8–14

If you are unsure of an answer during the question paper, then miss it out and come back to it at the end instead of wasting time thinking of a possible solution – but **always** come back to it. If you are still not sure, then use the data booklet to help, and at least have a sensible guess. If you guess, you may pick up some marks. If you leave it blank, you are guaranteed to lose all marks on this question.

DON'T FORGET
Remember: during the question paper, you will be given a data booklet. This contains all the calculations and data you will need. Don't forget to use this to help you!

ONLINE
If your teacher hasn't already given you a copy of the data booklet, it is worthwhile printing it out to help you answer your coursework. You can find a link to the data booklet on the BrightRED Digital Zone.

THINGS TO DO AND THINK ABOUT

When answering calculations in this course, make sure you are using **significant figures** within your final answers. In a number, all figures are significant, except zeros at the front (to show decimal points) or at the end. For example, in the number 1,234 there are 4 significant figures. In 0·0001234, there are also 4 significant figures.

The number of significant figures you use in your final answers should be equivalent to those used within the question. Information about this is also given at the front of your question paper.

ONLINE
Head to www.brightredbooks.net for a link to the official SQA past papers and answer schemes. It is highly recommended that throughout this course you visit these and try to answer the questions for the topic you are studying. One of the best ways to prepare for an exam is to answer exam questions!

ENGINEERING CONTEXTS AND CHALLENGES

ENGINEERING

ROLES AND DISCIPLINES

As you will already know from your studies in the National 5 course, a career in engineering is one of the most important and influential career paths anyone could take. There are many different branches and fields that exist.

Within the Higher course, you will focus on seven different yet intertwined disciplines.

Electrical Engineering

Although the name suggests that this discipline focuses purely on electrical systems, it involves much more than that. This is a branch of engineering that also includes power generation and distribution, motors and motor control, and the use of electromechanical devices.

Electronic Engineering

Although the name is similar to electrical engineering, and exam questions may refer to them as comparable disciplines, they have different roles that require different skills and knowledge.

Electronic engineering is about automatic control and the implementation of it. This means that electronic engineers may have the skills to design different electronic sensing circuits, be able to design and plan different programs for different functions, and then be able to write the code needed for it to run. To do this, they would also have the knowledge to understand how complex control systems work. This involves knowledge of components such as Op Amps, MOSFETs, relays and microcontrollers. Once the engineers have the system designed and working effectively, they must also have the skill to do energy audits to test that the system is working efficiently.

Civil Engineering

This branch of engineering deals with infrastructure. A civil engineer will plan, design and oversee the construction and maintenance of different facilities needed by modern civilisation. Civil engineers are involved in the design and planning of roads, railways, bridges, dams, irrigation projects, power plants and water and sewerage systems, and to do this they will assess the area of planned construction to ensure it is the most suitable location.

Structural Engineering

This branch of engineering is about analysing, designing, planning and researching structures, the materials they will be built from, and the forces that will act upon them to ensure that any given structure is safe and can support the needed loads. A structural engineer will have the knowledge and skills to understand and calculate the stability, strength and rigidity of a structure, ensuring that it will not collapse under different loads, forces or conditions.

Chemical Engineering

This branch of engineering deals with the chemical properties of materials and how these can be changed or altered. This could be a deliberate change, for example galvanising metal or coating a material with a certain paint to resist deterioration or corrosion; or knowing how different materials will act under certain conditions, such as exposure to weather conditions or elements.

Mechanical Engineering

This branch of engineering concerns itself with anything that moves. A mechanical engineer will analyse, design and develop mechanical devices to complete specific jobs. These can range from small components of a design, to extremely large plants and refineries, to machinery or vehicles.

Environmental Engineering

This is the branch of engineering that concerns itself with protecting all forms of life from adverse environmental effects such as pollution. Environmental engineers will have the knowledge of eco-friendly manufacturing methods to reduce the environmental impact of a project, and work with structural engineers to choose materials that are eco-friendly. They will also integrate low-carbon technologies in a project to reduce carbon footprints, and have the skills to adapt a design to limit its impact on the environment.

They also work on engineering solutions to improve recycling, water and air quality, waste disposal and management, and to improve public health.

Engineering Contexts and Challenges: Engineering

IMPACTS OF ENGINEERING

Within any engineering challenge or solution, we also have to consider the impacts that it will have on the population and the surrounding areas. These can be both positive and negative.

Social Impacts

- What will this do to the job market? Will it increase employment or training opportunities? Will jobs be lost due to its construction, e.g. by replacing a factory worker with a robotic system or by using driverless vehicles?
- What will this do for people's mental or physical health?
- Will there be improved infrastructure because of it?
- Will there be traffic disruption because of its creation?
- Will there be disturbances due to noise/pollution?
- Will it make things more accessible to people with disabilities?

Environmental Impacts

Things to consider in terms of environmental aspects are: could this provide habitats for wildlife or will there be a risk of damaging current animal habitats or ecosystems? Do schemes need to be identified to protect wildlife? Will there be a loss of green belt? Will nature be destroyed/damaged because of it? Will there be more demand on water or power services? What materials and manufacturing methods are used? Are they environmentally friendly or sustainable? Are there suitable waste-management systems and recycling procedures for the construction?

Economic Impacts

Things to consider in terms of economic impacts are: will this engineering solution bring money into the local area through tourism or other means? Will this employ more people, meaning more money being spent in the local area? Will it cost more money initially during its construction to meet certain legislation, e.g. to make it energy-efficient or by using energy-efficient materials? Will it create any tax breaks or benefit from any government incentives, e.g. due to lowering its carbon footprint? Will this attract other companies to invest in the area?

When answering questions about engineering impacts in the Higher exam paper, you have to be **very** specific. For example, stating 'driverless cars could cause jobs to be lost' would not be good enough if asked about the social impacts of autonomous vehicles. Instead, an explanation like 'driverless cars could cause loss of jobs in the taxi and public-transportation industry, as there would be no need for drivers, or within the long-distance haulage industry, as there would no longer be the need for drivers to take breaks' is what would be expected.

EMERGING TECHNOLOGIES

Another issue to consider when designing a solution is 'emerging technologies.' These are something that is being created and tested in Research and Development departments, but is not yet commercially viable. When examining a project, an engineer would have to consider several things: is something being created that would make the project more economical or efficient? Is there a new/emerging material or process that would be more suitable for this task? If it is powered, could it be driven in a more environmentally friendly way? Would using this technology change the world?

When describing any emerging technology, you must be thinking of cause and effect – what will this do, and what effect will this have on the world? For example, 'Artificial-leaf technology is currently being developed as a form of artificial photosynthesis to split water molecules into oxygen and hydrogen. This hydrogen can then be used to power transport and reduce the quantity of greenhouse gases we release into the atmosphere.'

THINGS TO DO AND THINK ABOUT

Research and create a datasheet on any emerging technology you can find. What is this technology, and how does it work? How will the development of this change the world?

DON'T FORGET

There is a big step up from National 5 to Higher, and your answers are expected to demonstrate this. Answers should never be a simple statement. A significant portion of the marks gained in the Higher question paper will be from descriptive answers. If you are asked about engineering roles in an exam, it should be of Higher English-level quality and should **describe**. This means you are talking about the **cause** of something, and the **effect** it then has.

DON'T FORGET

By searching online for 'Emerging Technologies', you will find lots of information out there – but, if you are struggling, *Scientific American* may be a good starting point, as well as the World Economic Forum, who create a top-ten list every year of emerging technologies.

DON'T FORGET

Remember – there is a difference between **knowledge** and **skills**, and you may be asked about these in the exam. A **skill** that an engineer has could be the ability to **design, calculate, simulate, analyse, justify** or **evaluate** part of an engineering project. **Knowledge** is what they need to know to allow this skill to be used.

ONLINE

Head to www.brightredbooks.net for additional rideos, online resorces, tests and more!

7

ENGINEERING CONTEXTS AND CHALLENGES
SUB-SYSTEM DIAGRAMS

OPEN-LOOP CONTROL

The 'Universal System Diagram' is a very basic way for working out how a piece of technology works.

To obtain a greater understanding of the system, though, we have to break it down into more detail. To do this, something called a 'sub-system diagram' must be drawn.

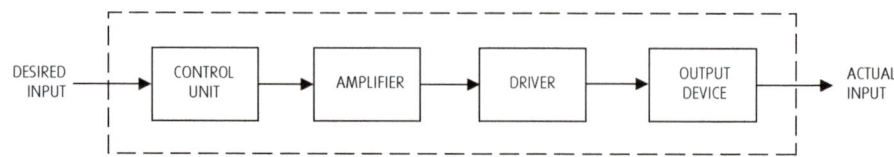

By using a sub-system diagram, it shows more detail of how a system works. For example, if we consider a hairdryer, a normal system diagram would only show what is going into and out of the system. Not much information can really be learned from this. By breaking it down into the sub-system diagram, it informs us of the internal workings and gives a clearer picture of how the system works.

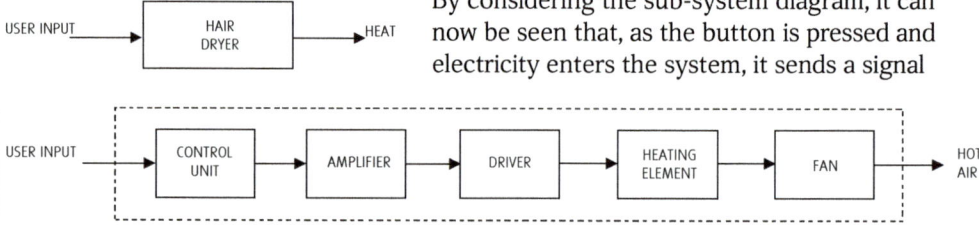

By considering the sub-system diagram, it can now be seen that, as the button is pressed and electricity enters the system, it sends a signal into the control unit to trigger the start of the process. The signal that this sends is now amplified by some form of amplifier, which in turn goes into a driver to increase the power to power the output transducer, in this case a heating element and fan. Hot air then leaves the hairdryer.

This is the simplest level at which a control system can process an input condition to produce a specified output. It is known as **open-loop control**. As you know from National 5, this is the most common form of control, and is used widely in domestic and industrial systems because it is cheap to install and simple to operate. It has a huge drawback, though – it has no way of sensing itself to see if the desired output is ever reached.

CLOSED-LOOP CONTROL

In closed-loop control, the value of the output is constantly monitored as the system operates, and this value is then compared with the desired (or reference) value. If a difference exists between the 'actual value' and the 'reference value', it realises that an error has occurred and will change the input into the system to reduce this output error to zero. Closed-loop control systems are capable of making decisions and adjusting their performance to suit any change in the output conditions. At National 5 you just drew a box and labelled it 'Control Unit', but at Higher you insert a component known as an 'error detector'.

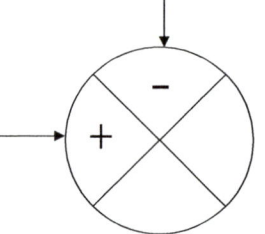

When drawing an error detector in a system diagram, you must **always** make sure the + is connected to the desired input, and the – is connected to the feedback loop. The other two quarters **must** be left blank.

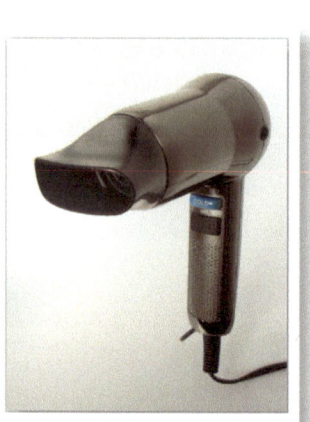

DON'T FORGET
The internal process aspects of the system should be drawn with a system boundary around it, like you learned in National 5.

DON'T FORGET
Electricity should never be seen as an input to a system unless it's an Energy Diagram. Instead, some form of user input our outside influence is more likely to be what initiates it.

DON'T FORGET
As you progress through the course, you will learn more about amplifiers. If it is known, write down the specific type of amplifier to be used.

VIDEO LINK
For a greater understanding of closed-loop control, watch the video on the BrightRED Digital Zone.

contd

A good example of a closed-loop system is a modern central heating system within a house. This will be designed to keep your house warm, but not so hot that it then becomes uncomfortable. It does this by allowing you to set the exact temperature you want your house to be at, and it will then change its input conditions depending on what the temperature is.

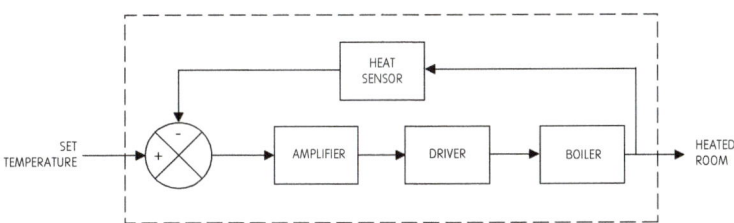

Within the sub system diagram, it can now be seen exactly how this system works.

- The desired temperature is input into the system.
- The error detector detects if the actual temperature and desired temperature differ, and it will generate a signal proportional to this difference (if it has a large difference it switches the heating on high, if the difference is small it switches the heating on a small amount).
- The signal is then amplified by the amplifier and fed to an output driver.
- The output driver will increase the power, allowing the boiler to be heated.
- The boiler switches on and heats the room.
- At the same time, it sends a signal back to the error detector, letting it know if a difference exists between the actual temperature and the desired temperature.
- As long as there is a difference between them, the system will attempt to reduce this difference – if the room is too cold, it will increase the signal and therefore the outgoing temperature. If the room is too hot, it will switch off the heating until the temperature drops.

DON'T FORGET

Within the feedback sensor, it might be a good idea just to write down whatever type of sensor it is – e.g. speed sensor or temperature sensor, instead of accelerometer or thermistor. If you write the wrong thing, e.g. a speedometer as a 'speed sensor', you will lose that mark in the exam, even if it is clear what you meant.

THINGS TO DO AND THINK ABOUT

A street-lighting system is controlled automatically. When the outside light drops below a set level, a lamp comes on.

a) Draw and complete this diagram.
b) State the name of this type of control.

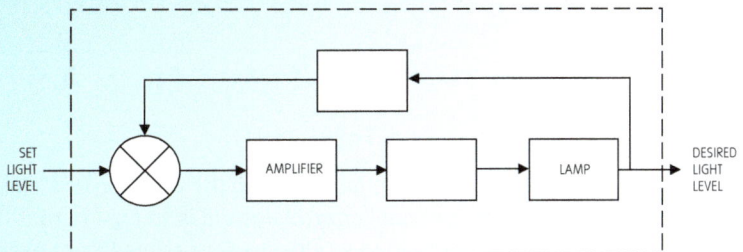

c) Describe how this circuit works.

ONLINE TEST

Test your knowledge of this topic at www.brightredbooks.net

ENGINEERING CONTEXTS AND CHALLENGES
ANALOGUE CONTROL SYSTEMS

TWO-STATE CONTROL

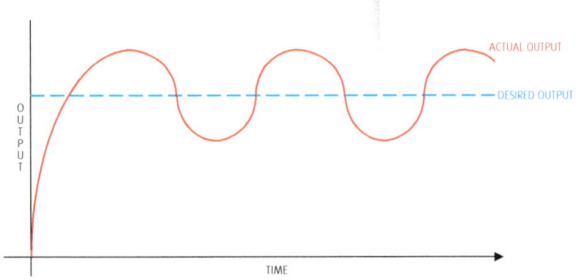

Many control systems have to process analogue signals such as heat, light and movement. Therefore analogue closed-loop control systems require an analogue processing device such as the **operational amplifier** ('Op Amp'). This is a component we will delve further into within the 'Electronics and Control' chapter of the book.

One of the most common control applications involves using the Op Amp as a **comparator**. In its simplest form, an Op Amp acting as a comparator just compares two voltage signals, V_1 and V_2, if V_1 is the feedback and V_2 is the **reference voltage** for it to compare against. If V_1 is higher than V_2, then the output, V_{OUT}, is 'low'. If V_1 is lower than V_2, the output is 'high'.

If the output can only have two options (either 'high' or 'low'), then any system using an Op Amp as a comparator is known as a **two-state closed-loop control system**.

The major drawback of a two-state closed-loop control system is that the desired output is rarely met. Because it essentially acts in a digital manner, meaning the system is either on or off, it can never fully meet the desired output, as it will always be just above or just below it.

PROPORTIONAL CONTROL

Proportional control is another way of controlling the output, without having the same negative traits. This is a system that reduces the error **proportionally** – if the error is large, it sends a large negative signal to fix the problem. If the error is small, it sends a small signal. This allows the desired output to eventually be reached.

This is usually achieved with a type of Op Amp known as a **difference amplifier**, which we will examine in more detail in the 'Electronics and Control' chapter of the book.

An Op Amp set up as a **difference amplifier** amplifies the difference between the **reference level** and **feedback signal** – essentially, it is looking at the difference that exists between the two signals, which allows its output to be proportional to this difference. This helps to prevent the 'overshoot' and 'time lags' that can be seen in a comparator-based system. Because of this it allows the desired output to be reached, as the difference between the two signals will eventually be 0.

TYPES OF FEEDBACK IN A SYSTEM

Negative Feedback

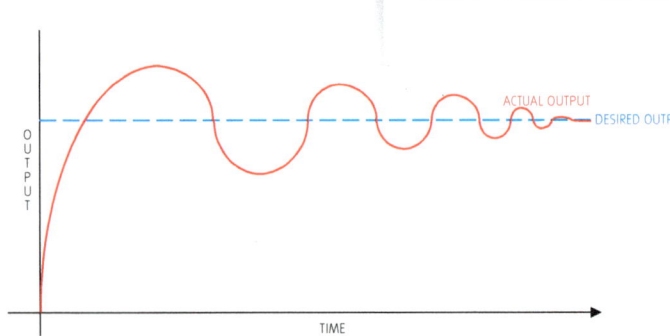

As you have just learned, the purpose of a closed-loop proportional control system is to ensure that the output is maintained as closely as possible (if not exactly) to the desired output level.

If we take the example of a central heating system, a graph of the temperature in a room might appear like this.

This control system has a 'lag'. This means that the system is constantly trying to pull the temperature of the room back towards the set temperature level by reducing the error. This type of control uses **negative feedback** to reduce the error.

contd

Engineering Contexts and Challenges: Analogue Control Systems

Positive Feedback

The opposite effect can be created by reinforcing the error. An example of this in use is when a microphone is held too close to an audio speaker.

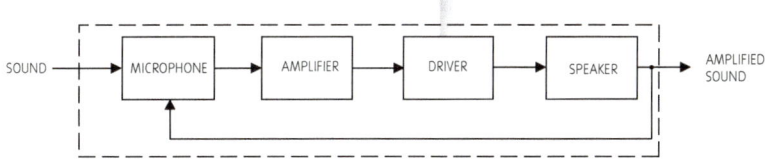

When this happens, the sound is picked up by the microphone, amplified, and then output through the speaker.

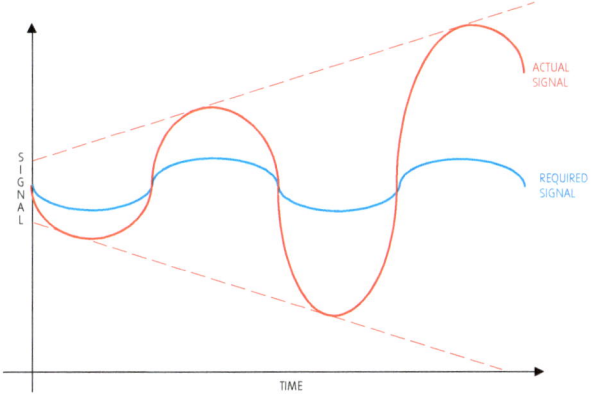

This amplified sound is then picked up by the microphone, re-amplified and continued in an endless loop. The end result is a high-pitched sound, which is represented by this graph. Although positive feedback does have some very useful applications, negative feedback is far more widely used in control systems.

ONLINE TEST

Test your knowledge of this topic at www.brightredbooks.net

DON'T FORGET

A significant proportion of the marks at Higher come from descriptive answers to questions. Make sure you totally understand these systems and how they work to allow you to answer in detail.

THINGS TO DO AND THINK ABOUT

An outdoor swimming pool uses a two-state closed-loop control system to regulate the water temperature.

a) For the pool's temperature-control system, draw the sub-system diagram.

b) State the name of the operational amplifier (Op Amp) configuration that would be used in this type of control.

c) Sketch a graph of temperature against time, showing the desired temperature, and show how the temperature of the water changes as it is heated from a lower temperature.

d) The control system is replaced by a closed-loop proportional control system. Describe how proportional control would be used to maintain a steady temperature.

11

ENGINEERING CONTEXTS AND CHALLENGES

ENERGY AND EFFICIENCY

OVERVIEW

Within a system, it is important that an engineer knows the energy going into and out of the system so they know how efficient it is. They can then discover ways to make it more efficient, or redesign it to be a more effective system. By making products and processes more energy-efficient, it not only helps to reduce the amount of waste energy (therefore saving money), but it will also help to reduce the amount of greenhouse gases that are emitted through the creation and use of the system. This not only ensures that the system has a lower carbon footprint, but also it helps to ensure that the system is running as well as it should be.

In these energy-loss diagrams, you can see the need for the reduction in use of the old-style filament light bulbs. Being only 10% efficient means that 90% of the energy is being transferred into non-useful forms, such as heat – and nobody has said 'it's getting cold, let's switch on the lights'! This meant that engineers had to think of ways to alter or redesign this everyday object to lower its carbon footprint, make it more efficient, and make it more suitable for purpose.

In recent years, the use and development of LED bulbs in a domestic environment has also dramatically increased, as these are even more efficient. Although the initial purchase cost of LED bulbs in comparison to others is higher, the lifespan of the bulb is longer, meaning that fewer purchases of light bulbs are needed, and the running energy costs are reduced due to their high efficiency. This also means that energy consumption is less, and less wastage is created due to not having to dispose of light bulbs as often, thereby reducing the carbon footprint that is left.

Before energy is calculated, sometimes different calculations will need to be done first to allow you to obtain the information needed.

Work Done

Work results when a force acting upon an object causes a motion or displacement.

To calculate work done, the following equation must be used: $E_w = Fd$

(work done = force applied × distance moved)

Weight

People often confuse mass and weight. The **mass** of an object is the amount of matter it contains, whereas **weight** is the force caused by gravity.

To calculate **weight**, the following equation must be used: $W = mg$

(Weight = mass × gravity)

Power

Power is defined as the rate at which work is done upon an object. To calculate power, the following equation must be used: $P = E/t$

(Power = energy transfer ÷ time)

Once you have converted the information given into units you can use, you should be able to use the equations relating to energy. These are no different from the ones you learned in National 5.

Electrical Energy

To calculate electrical energy, the following equation must be used: $E_e = VIt$

(Electrical energy = voltage × current × time)

DON'T FORGET

Within National 5, you learned several different calculations – and these will be needed again in Higher (these can be easily revisited by reading the Energy Calculations section within the BrightRED Study Guide for National 5 Engineering Science, where in-depth exemplars are also given). The difference in the Higher course, though, is that there is more emphasis on manipulating and combining the formulae to obtain answers.

DON'T FORGET

Don't think you can now forget everything you learned in National 5. The Higher course covers all the same topics, but now the difficulty will be increased, and it will go into far greater detail.

contd

12

Engineering Contexts and Challenges: Energy and Efficiency

Kinetic Energy

To calculate kinetic energy, the following equation must be used: $E_K = \tfrac{1}{2}mv^2$

(Kinetic energy = 0·5 × mass × velocity²)

Potential Energy

To calculate potential energy, the following equation must be used: $E_p = mgh$

(Potential energy = mass × gravity × height)

Heat Energy

To calculate heat energy, the following equation must be used: $E_h = cm\Delta T$

(Heat energy = heat capacity of material × mass × change in temperature)

Elastic Strain Energy

Strain energy is a new energy you will not have covered in the National 5 course. This is seen as similar to potential energy, as it is a stored energy, but it differs in that strain energy is stored in a body because it has been stretched or compressed.

To calculate elastic strain energy, the following equation must be used: $E_s = \tfrac{1}{2}Fx$

(Strain energy = 0·5 × force × distance of compression or extension)

EFFICIENCY OF A SYSTEM

Energy cannot be destroyed or created; it can only be transformed from one form into another. Hence, you also should know that the energy output from a system must be equal to its input. Unfortunately though, some of that energy does not transfer into the energy that you want. When an energy conversion takes place, there is always an unwanted energy change – usually in the form of heat or sound through friction within any moving parts.

It is possible to look at how well an energy system is operating by calculating its efficiency. The efficiency of an energy transformation is a measure of how much of the input energy appears as useful output energy. This then allows an engineer to evaluate the system and see how it could be improved.

To calculate efficiency, one of the following equations must be used:

$\eta = (E_{out}/E_{in}) \times 100\%$ <u>or</u> $(P_{out}/P_{in}) \times 100\%$

Efficiency = (useful energy out ÷ useful energy in) × 100%

or

Efficiency = (useful power out ÷ total power in) × 100%.

THINGS TO DO AND THINK ABOUT

When an engineer tests a 12 V, 7·5 A car-tyre compressor pump, they find that it produces 430 J of energy over a time period of 15 seconds. Work out the input energy for the system, and the efficiency of the system, and then draw your results in the form of an energy-audit diagram.

DON'T FORGET

The gravitational pull on the earth is 9·8 ms⁻². This info is in your data booklet – so, if you forget, look there!

DON'T FORGET

Don't forget to transfer time into seconds.

ONLINE TEST

Test your knowledge of this topic at www.brightredbooks.net

DON'T FORGET

The specific heat capacity of water is 4,200 Jkg⁻¹K⁻¹, and this is different from what you used in National 5, as it's now put to 2 sig figs. Remember always to check your data booklet.

DON'T FORGET

All these calculations were used in National 5. For a greater understanding of how they are used, go back and read the 'Energy Calculations' section in the BrightRED Study Guide for National 5 Engineering Science, where in-depth exemplars are given.

ELECTRONICS AND CONTROL

ANALOGUE ELECTRONICS: ELECTRONIC CIRCUITS

THE BASICS OF ELECTRONIC CIRCUITS

As you know from your previous studies, an electronic circuit is made up of a collection of electrical components that are connected together in a closed-loop circuit designed to do a specific task.

When describing what happens in a circuit, we often refer to the voltage, the current and the resistance. To explain how this works, a common analogy that is used is a water tank. The voltage is the water pressure, and the current is the water flow. Consider a water tank held at a height above the ground, and at the bottom of this tank there is a hose.

The pressure at the end of the hose would represent the voltage, and the water in the tank would represent the charge. The more water that is in the tank, the higher the charge, and therefore the greater pressure that occurs at the end of the hose. Think of this tank like a battery: if we drain the tank a little, the pressure created at the end of the hose will go down. This is similar to what happens in a torch when the batteries are low – the torch becomes dimmer.

There would also be a decrease in the amount of water that would flow through the hose. This reduction in pressure would lead to less water flowing. The amount of water that is flowing through the hose is like electrical current. The higher the pressure, the higher the flow, and vice versa.

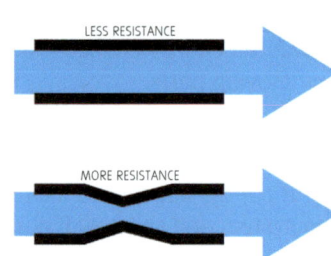

The third factor to be considered in this analogy is the width of the hose. The width of the hose is like the resistance in a circuit. It stands to reason that we can't fit as much through a narrow pipe as we could through a wider one. A narrower pipe would 'resist' the flow of water.

Ohm's Law

VIDEO LINK

Go to www.brightredbooks.net to watch a video to strengthen your grasp of Ohm's Law.

Ohm's Law is the relationship in a circuit between the voltage, the current and the resistance.

V = IR

(Voltage = current × resistance)

Power in a Circuit

Electrical power is the rate at which energy is used within the circuit.

P = IV

(Power = current × voltage)

or

P = I²R

(Power = (current × current) × resistance)

or

P = V² / R

(Power = (voltage × voltage) × resistance)

DON'T FORGET

All these calculations were used in National 5. For a greater understanding of how they are used, go back and read the 'Analogue Electronics' section in the BrightRED Study Guide for National 5 Engineering Science, where in-depth exemplars are given.

Electronics and Control: Analogue Electronics: Electronic Circuits

SERIES CIRCUITS

A series circuit is the simplest form of circuit, but it has a major disadvantage: if one of the components breaks, the whole circuit stops working.

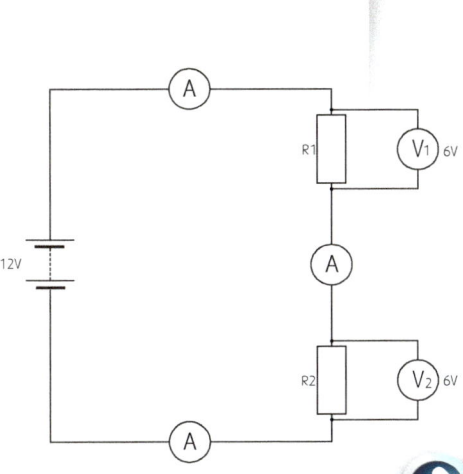

Voltage in a Series Circuit

The voltage is divided up between all the components in a series circuit.

$V_{cc} = V_1 + V_2 \ldots$

You can see in this circuit that this is true, as two voltmeters in the circuit add up to the total voltage entering the circuit through the battery.

Current in a Series Circuit

The current flows through all the components equally in a series circuit. It does not matter where you put the ammeter, as it will always show the same amount of amps.

Resistance in a Series Circuit

To calculate the resistance in a series circuit, you have to add up the total resistance of each component.

$R_{total} = R_1 + R_2 \ldots$

ONLINE

Visit the BrightRED Digital Zone to strengthen your knowledge and understanding of series and parallel circuits.

PARALLEL CIRCUITS

A parallel circuit differs from a series one, in that it is broken up into different branches. This allows the current to flow through each branch separately. If one component breaks, only that branch stops working.

Voltage in a Parallel Circuit

Within a parallel circuit, each branch receives the supply voltage.

Current in a Parallel Circuit

When the current in a parallel circuit reaches a junction, it splits the current up, with some going along one branch, and the rest going along another.

$I_{total} = I_1 + I_2 \ldots$

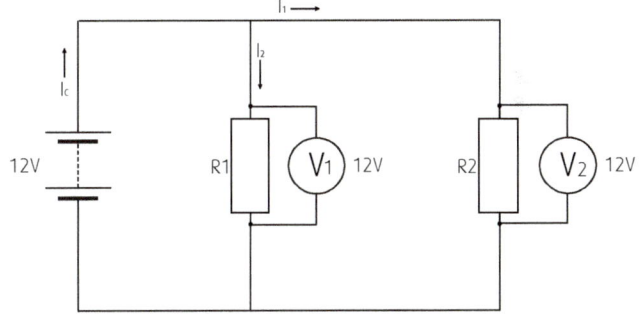

Resistance in a Parallel Circuit

To calculate the total resistance in a parallel circuit, there are two equations we can use:

$R_t = (R_1 \times R_2) / R_1 + R_2$

(this one can only be used with two resistors in parallel)

or

$1 / R_t = (1 / R_1) + (1 / R_2) \ldots$

(this **must** be used if three or more resistors are in parallel)

ONLINE TEST

Visit www.brightredbooks.net to test your knowledge of series and parallel circuits.

THINGS TO DO AND THINK ABOUT

Read over your National 5 notes and answer the Analogue Electronics questions from the SQA past-papers website. This is the basis of all analogue electronics. If you don't fully understand this, you will struggle to grasp the more complex material within the Higher course.

ELECTRONICS AND CONTROL

ANALOGUE ELECTRONICS: VOLTAGE-DIVIDER CIRCUITS

OVERVIEW

An input transducer is used to change a physical parameter into an electrical signal as the physical conditions change. These are some variation of a resistor – for example, an LDR or a thermistor, which vary their resistance as the light or temperature respectively change. To do anything meaningful, we then have to convert this resistance change into a voltage change, which is done using a voltage-divider circuit. A typical voltage-divider circuit is shown here.

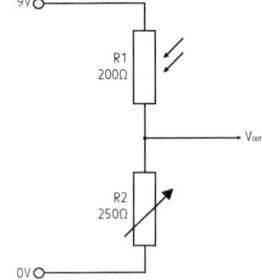

To work out the output in this type of circuit, we use the following equation:

$V_{out} = (R_2 / (R_1 + R_2)) \times V_{cc}$

In this example, V_{out} would be calculated by:

$V_{out} = (R_2 / (R_1 + R_2)) \times V_{cc}$

$= (250 / (200 + 250)) \times 9$

$= 0.556 \times 9$

$= \underline{5\ V}$

Extended Voltage-Dividers

Within the Higher course, it is sometimes possible to have a voltage-divider circuit that has more than two resistors. This looks complicated at first, but all you have to do is remember your calculation for resistors in series.

In this case, several of the resistors would be combined together and be seen as having one total resistance. In this circuit, when working out V_{out} A, resistors 2, 3 and 4 would be added together and seen as R2 in the Voltage Divider Equation.

For example, if V_{out} C was to be calculated, then R_1, R_2 and R_3 would be seen as in series with each other and be summed together to create one overall total resistance.

$V_{out} = (R_2 / (R_1 + R_2)) \times V_{cc}$

$= (250 / (100 + 200 + 150) + 250) \times 9$

$= 0.357 \times 9$

$= \underline{3.2\ V}$

Missing Information

Depending on the information given, it isn't always possible to determine the resistance using this calculation. If the value one of the resistors is unknown, this is impossible. If the voltage going across at least one of the components is known, though, the resistance can be calculated using this equation:

$(V_1 / V_2) = (R_1 / R_2)$

VIDEO LINK

Head to www.brightredbooks.net and watch the video on voltage-dividers to strengthen your knowledge and understanding of them.

TRANSDUCER GRAPHS

As previously stated, a voltage-divider circuit uses a component like an LDR or a thermistor to change light and temperature respectively into resistance, essentially making a light or heat sensor (or, if the positions of the two components are swapped around, a dark or cold sensor).

Within the Higher course, you may need to know the resistance of one, but are only given the lux (light level) or the temperature. This is easily determined, though, by using a thermistor or LDR graph – and a copy of this graph will **always** be given in the question.

LDR Graph

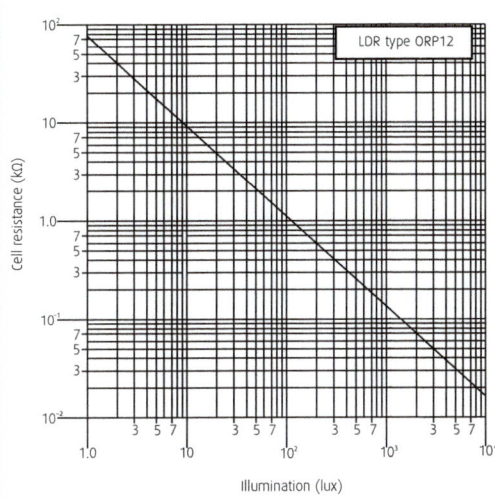

Thermistor Graph

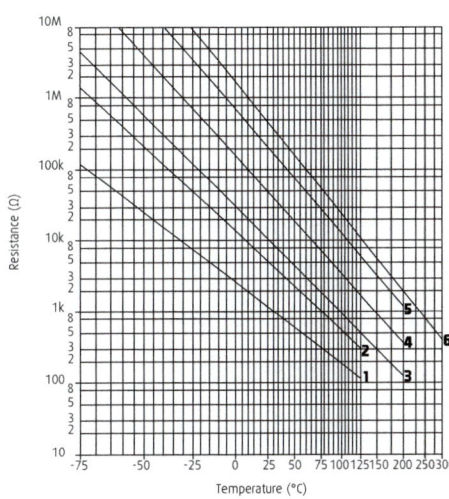

It is important to note that the resistance and the lux/temperature go up in different increments, so make sure you are reading it properly. For example, on the thermistor graph, the resistance moves up in different decimal prefixes. It starts at 10, then as it moves up it shows 2, 3, 5 and 8, then on to 100. In this case, the 2 represents 20, the 3 represents 30, and so on. As it moves into the 100s, the next 2 represents 200, the 3 is 300, and it continues in this fashion.

Example:

To work out the resistance in this graph (or the temperature/lux), it is as simple as finding the information you are given, following the line until it hits the diagonal line representing that specific output transducer, and bouncing it along or down. For example, to calculate the temperature of a type 1 thermistor at 5 kΩ, you first have to find the resistance. Follow the grid until it hits the line representing that specific thermistor, and follow it down to the temperature. In this case, the temperature would be −15°C.

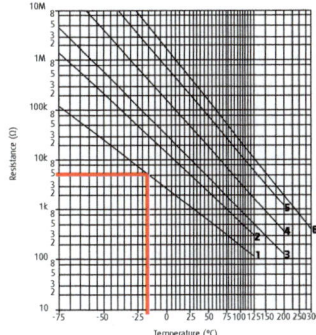

THINGS TO DO AND THINK ABOUT

This diagram shows the input sub-system for a complex electronic circuit. The output will switch on when the light level drops to 200 lux and V_{out} is 2·9 V.

Calculate the required value of R.

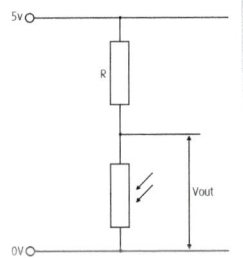

ONLINE TEST

Test your knowledge of this topic at www.brightredbooks.net

DON'T FORGET

This is also covered in the National 5 course, so remember to look back at your old notes and the National 5 Engineering Science Study Guide to help your understanding.

DON'T FORGET

If you find the line hard to follow, feel free to use a ruler and draw on the graphs if you come across this in an exam situation. That way, you can ensure you are getting a straight line.

ELECTRONICS AND CONTROL

ANALOGUE ELECTRONICS: OPERATIONAL AMPLIFIERS 1

ONLINE

Head to the BrightRED Digital Zone for additional videos, links, tests and more!

USES AND CHARACTERISTICS OF OP AMPS

As has been illustrated in the previous pages, many control systems involve processing analogue signals such as heat, light, pressure or movement. This means that analogue closed-loop control systems such as these will therefore require an analogue processing device. The most commonly used one is the **Operational Amplifier** (Op Amp).

An Op Amp can be used to add, subtract, multiply, divide, integrate and differentiate electrical voltages. It can also amplify both DC and AC signals.

An ideal amplifier should have the following qualities:

- an infinite input resistance, so that very little current is drawn from the source (V = IR)
- zero output resistance
- an extremely high gain
- no output when the output is 0.

THE COMPARATOR

As you can see in the Op Amp symbol, there are two input signals and one output to the device. In addition to this, though, you may notice that it has its own external power supply to both positive and negative voltages.

It must be noted, though, that the output from an Op Amp **cannot** be greater than 85% of its supply voltage. For example, if an Op Amp has a +10 V and −10 V supply, then the output from it must lie somewhere between +8·5 V and −8·5 V.

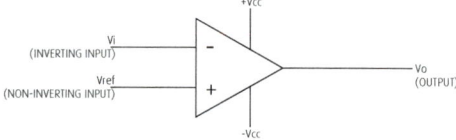

In a simple configuration such as this, the Op Amp will act as a **comparator**. This means that the Op Amp is designed to compare two separate voltage signals, in this case V_i and V_{ref}. V_i is the input signal to the Op Amp, which is coming from the input transducer, and V_{ref} is the reference voltage that it will use for a comparison. Within a comparator, if V_i is higher than V_{ref}, the output (V_o) is 'low'. This means that it will have no output. If V_i is lower than V_{ref}, the output will be 'high'. This means that the Op Amp will create an output.

DON'T FORGET

You may be asked how to modify a circuit like this so that the green LED comes on when it is dark. This can be done in one of two ways: either swap the inverting and non-inverting inputs over, or swap the positions of R1 and R2 in the voltage-divider leading to V_i.

Example:

In this example, an Op Amp is set up to compare the light level (the voltage-divider that consists of R1 and R2) against the reference level (the voltage-divider of R3 and R4).

If the LDR has a high resistance, meaning that its voltage-divider output (and the non-inverting input of the Op Amp) is higher than the inverting reference input, then the output of the Op Amp will be high. In this example, the V_i will be high and will create an output of 10·2 V because this figure is 85% of the Op Amp supply voltage. This, in turn, will switch on the green LED.

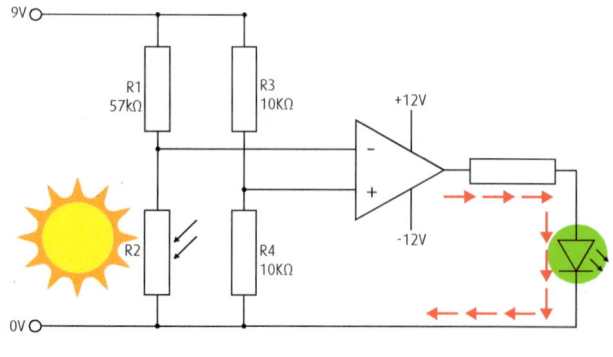

contd

Electronics and Control: Analogue Electronics: Operational Amplifiers 1

If the LDR has a low resistance due to a lack of light, the voltage output of this voltage-divider, and hence the input into the non-inverting Op Amp, is low. This means that the reference input voltage will be greater than the input signal, so the Op Amp will low and give out 0 V. No voltage will go through the Op Amp, and the LED will therefore be off.

The major disadvantage of this configuration is that it works to produce a two-state control. This means that it works in a digital fashion and must be either on or off, with no in-between. This means the actual output will either be just above, or just below, the desired output.

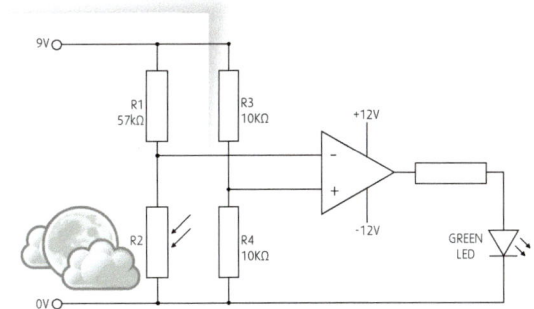

THE DIFFERENCE AMPLIFIER

Within the **difference amplifier** configuration, both inputs are used, and the Op Amp amplifies the **difference** between the two signals.

If there is no difference between the inputs, the output will be 0. If a difference exists, the output will switch on, and the amount the signal is amplified by is purely dependent on how big that difference is. A difference amplifier is used to produce proportional control. If a large difference exists, the output signal will therefore be large. If the difference is small, then the output signal will be small.

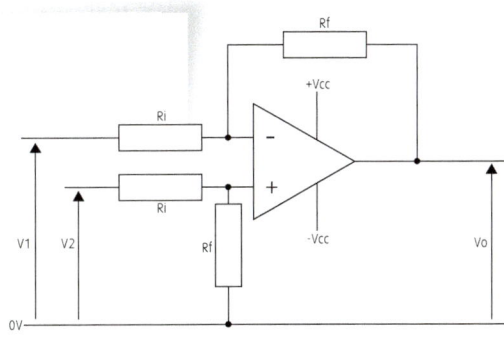

To calculate the output voltage of the circuit, we can use the following equation:

$V_o = (R_f / R_i) \times (V_2 - V_1)$

To calculate the gain (Av) of the Op Amp, there are two calculations you can use for a difference amplifier:

$Av = V_o / (V_2 - V_1)$ **or** $Av = R_f / R_i$

Example:

This example shows a difference amplifier connected to two voltage-dividers that have strain gauges on them to test the pressure put on a beam. (A strain gauge is essentially a sensor whose resistance varies with applied force, pressure or tension.)

Using the calculations stated, it can be proved that if there is no difference between the inputs, the output will be 0.

By using our previous knowledge, we can firstly realise that if no strain exists, then RG1 must equal RG2. Therefore, if we can calculate V_{out} on one of these voltage-dividers, it will equal the same V_{out} on the other voltage-divider.

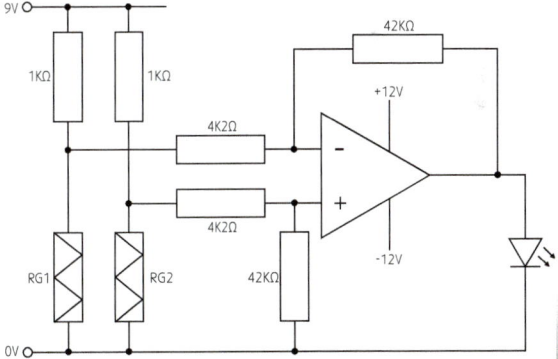

If we say that, when not under strain, RG1 = RG2 = 275 Ω,
$V_{out} = (R_2 / R_1 + R_2) \times V_{cc} = (275 / (1{,}000 + 275)) \times 9$
$V_{out} = \underline{\mathbf{1 \cdot 9\ V}}$
$Av = R_f / R_1 = 42 / 4 \cdot 2$
     = **10**
$V_o = (A_v) \times (V_2 - V_1) = 10 \times (1 \cdot 9 - 1 \cdot 9)$
$V_{out} = \underline{\mathbf{0\ V}}$ Hence, the Op Amp will be off.

It can also be proved that if a difference exists, the output will switch on. If we change it so that there is an increased force put on RG2 so that this is value is now 235 Ω.

$V_{out} = (R_2 / R_1 + R_2) \times V_{cc} = (325 / (1{,}000 + 325)) \times 9$
$V_{out} = \underline{\mathbf{2 \cdot 2\ V}}$
$V_o = (Av) \times (V_2 - V_1) = 10 (2 \cdot 2 - 1 \cdot 9)$
$V_{out} = \underline{\mathbf{3\ V}}$ Hence, the Op Amp will be on.

DON'T FORGET

These calculations and pictures of the configurations are in your data booklet, so there is no need to memorise these. Just remember to use your data booklet!

DON'T FORGET

Remember: when calculating, you should be using the correct amount of significant figures. On the front of the exam paper, it gives advice on how these should be applied. As a general rule, for final answers you should use the same amount of figures that are used in the question; but usually answers that have 2 more or 1 less will be accepted.

THINGS TO DO AND THINK ABOUT

Simulate the circuits shown in the examples, using an electronic simulation package such as Yenka to verify that the theory you are learning proves to be true.

ELECTRONICS AND CONTROL

ANALOGUE ELECTRONICS: OPERATIONAL AMPLIFIERS 2

> **ONLINE**
>
> Head to www.brightredbooks.net for additional activities, videos and tests on this topic!

THE INVERTING AMPLIFIER

The **inverting amplifier** is a special type of Op Amp configuration that produces an output that is exactly 180 degrees out of phase to its input signal. This means it produces **negative feedback** across the op-amp circuit, which essentially means that if the signal going into the Op Amp is positive, it will leave as an amplified negative signal, and vice versa.

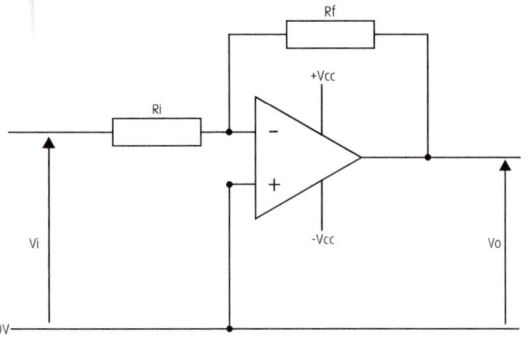

Within this type of configuration, the input signal is connected to the inverting/negative terminal via the input resistor Ri, and the non-inverting/positive terminal is connected to 0v. Another resistor (Rf) is used to connect the output terminal to the inverting/negative input terminal which creates negative feedback within the system. This negative feedback sends a fraction of the output signal back to the input and, in effect, produces a **closed loop** circuit to the amplifier. The feedback connection between the output and the inverting input terminal forces the differential input voltage towards zero and uses this to accurately control the overall gain of the amplifier, as well as stabilizing the output.

In many examples you will see throughout this course, a circuit that includes an inverting Op Amp usually includes two of them – one directly following the other. This is done to amplify the signal twice, but also to allow the inverted signal to be re-inverted back into a positive value.

To calculate the output voltage of an inverting Op Amp circuit, we can use the following equation:

$$V_o = (R_f / R_i) \times V_i$$

To calculate the gain (Av) of the Op Amp, there are two calculations you can use for an inverting amplifier:

$$A_v = V_o / V_i \quad \underline{\text{or}} \quad A_v = - R_f/R_i$$

> **DON'T FORGET**
>
> Don't be scared off by a complex-looking circuit. Break it down into small chunks, use the data booklet to find out the calculation you need for the end goal, then continue to use your data booklet and knowledge to work out how to obtain the different aspects of that calculation.

Example:

This example shows a circuit that utilises an inverting Op Amp. By applying your acquired knowledge, several different aspects of this circuit can be calculated. This is probably similar to what would be asked in an assessment, but it won't be as easy as working out a simple calculation. Instead, it is to be expected that you will have to work out several different aspects of the circuit before you can reach your end goal.

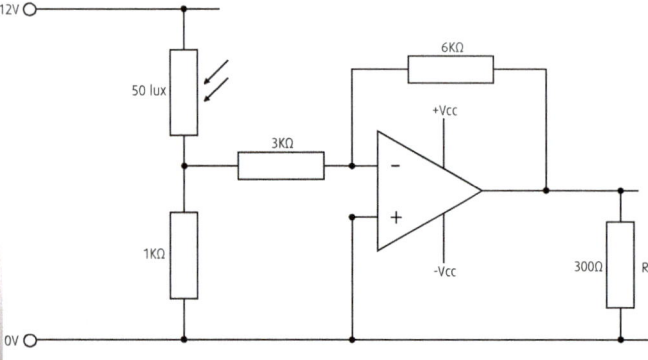

For example, the resistance of the LDR can be calculated using the LDR graph, as you know the light level (lux):

50 lux = 2 kΩ

Knowing this, V_{out} from the voltage-divider can easily be calculated:
$V_{out} = (R_2 / R_1 + R_2) \times V_{cc} = (1 / (2 + 1)) \times 12$
= **4 V**

The gain of the Op Amp can also be calculated:
$A_v = -R_f / R_i = -6 / 3$
= **-2**

With this information, the output of the Op Amp can now also be calculated:

contd

$A_v = V_o / V_i$
$\Rightarrow V_o = A_v \times V_i = -2 \times 4$
$= \underline{\mathbf{-8\ V}}$

And, now that this has been found out, the current going through the load resistor can be calculated:
$V = IR$
$\Rightarrow I = V / R = -8 / 300$
$= \underline{\mathbf{-26 \cdot 7\ mA}}$

THE NON-INVERTING AMPLIFIER

In this configuration of an Op Amp, the input voltage signal (V_i) is applied directly to the non-inverting (+) input terminal. This means that the output gain of the amplifier becomes positive, which is in contrast to the inverting-amplifier circuit in the previous pages. Like the inverting amplifier, it is used to multiply the input voltage by the gain, but in this case it **does not** invert the signal.

To calculate the output voltage of the circuit, we can use the following equation:

$$V_o = (1 + (R_f / R_i)) \times V_i$$

To calculate the gain (Av) of the Op Amp, there are two calculations you can use for a non-inverting amplifier:

$Av = V_o / V_i$ **or** $Av = 1 + (R_f / R_i)$

Example:

This example shows a circuit that utilises a non-inverting Op Amp. Like before, several different aspects of this circuit can be calculated by applying the knowledge you have gained.
This circuit uses a type 1 thermistor in a voltage-divider circuit to create a temperature sensor. It has been set so that when it is warm it shows a temperature of 30°C, and in the cold it shows 10°C.

Firstly, we can calculate the gain of the Op Amp:

$Av = 1 + R_f / R_i = 1 + (20 / 40)$
$= \underline{\mathbf{1 \cdot 5}}$

We can also determine the voltages that would appear on the voltmeter in both light conditions:

Case 1: Thermistor = 10°C

Using the thermistor graph, we can find out that 10°C = 2 KΩ.

$V_{out} = (R_2 / R_1 + R_2) \times V_{cc}$
$= (10 / (2 + 10)) \times 9$
$= \mathbf{7 \cdot 5\ V\ (Vi)}$

V_{out} (from Op Amp) $= Av \times Vi$
$= 1 \cdot 5 \times 7 \cdot 5$
$= \underline{\mathbf{11 \cdot 25\ V}}$

Case 2: Thermistor = 30°C

Using the thermistor graph, we can find out that 30°C = 1 KΩ.

$V_{out} = (R_2 / R_1 + R_2) \times V_{cc}$
$= (10 / (1 + 10)) \times 9$
$= \mathbf{8 \cdot 2\ V\ (Vi)}$

V_{out} (from Op Amp) $= Av \times Vi$
$= 1 \cdot 5 \times 8 \cdot 2$
$= \underline{\mathbf{12 \cdot 3\ V}}$

> **DON'T FORGET**
>
> If asked to draw an Op Amp configuration in any form of assessment, it must be **exactly** the same as it is in the data booklet. This means drawing **all** labels and information shown there.

> **DON'T FORGET**
>
> It is easy to mix up the non-inverting and the inverting Op Amp, so make sure you are taking care to look properly at its connections, and **ALWAYS** use your data booklet to double check and be sure!

> **DON'T FORGET**
>
> Gain does not have a unit.

THINGS TO DO AND THINK ABOUT

This circuit is used to amplify the signal produced by a flow sensor in part of a water-treatment plant.

a) Explain the reasons for the inclusion of the second Op Amp in this circuit.

b) Calculate the required value of resistor R, so that V_{out} reaches its maximum value when V_{in} is 0·9 V.

ELECTRONICS AND CONTROL

ANALOGUE ELECTRONICS: OPERATIONAL AMPLIFIERS 3

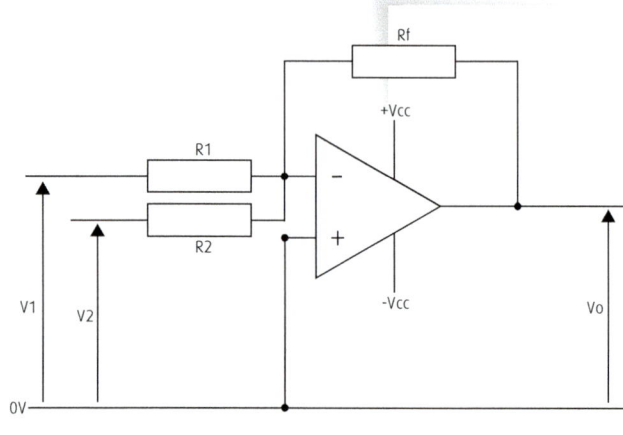

THE SUMMING AMPLIFIER

The **summing amplifier** is a special type of operational-amplifier circuit that is used to combine the voltages of two or more inputs into one single output.

Summing Op Amps are used in a wide range of electronic circuits – for example, on an audio mixer in a recording studio. The summing amplifier adds the waveforms together from various channels (e.g. vocals, guitar, drums etc.) before sending the mixed signal to a recorder.

To calculate the output voltage of the circuit, we can use the following equations:

$$V_o = (Av_1 \times V_1) + (Av_2 \times V_2) + \ldots \quad \text{or} \quad V_o = -R_f(V_1/R_1 + V_2/R_2 \ldots)$$

To calculate the gain (Av) of the Op Amp, we have to calculate what it would be for each individual input:

$$Av_1 = -R_f/R_1 \qquad Av_2 = -R_f/R_2 \qquad Av_n = -R_f/R_n$$

VIDEO LINK

Visit the BrightRED Digital Zone to learn more about how a summing Op Amp works.

Example:

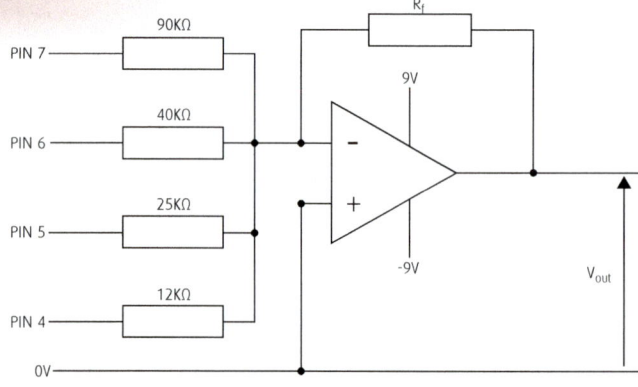

This example shows a microcontroller-based control system that ensures that fuel is delivered at a constant rate when aircraft are being refuelled. The microcontroller outputs are processed by the summing amplifier, which controls the pump speed.

The microcontroller output pins are at 5 V when switched on.

For the diagram shown:

a) Calculate the value of resistor R_f so that the maximum output voltage V_{out} is −6·8 V.
$V_{out} = -R_f(V_1/R_1 + V_2/R_2 + V_3/R_3 + V_4/R_4)$
$-6·8 = -R_f(5/90 + 5/40 + 5/25 + 5/12)$
$6·8 = R_f \times 0·798$
$R_f = \underline{\mathbf{8·5\ k\Omega}}$

b) Calculate the output voltage V_{out} when only pins 7 and 5 are sending out an output voltage.
$V_{out} = -R_f(V_2/R_2 + V_3/R_3)$
$V_{out} = -8·5(5/40 + 5/25)$
$V_{out} = \underline{\mathbf{-2·8\ V\ (2\ Sig\ Figs)}}$

DON'T FORGET

Remember to use significant figures in your answer. Rule of thumb is that the number of significant figures used in a final answer should be equivalent to the least significant data value given in the question, but answers that have two more, or one less, will be accepted. Information on this will be given on the front of your exam paper so be sure to read it carefully!

THE VOLTAGE-FOLLOWER

A **voltage-follower** is an Op Amp configuration which **always** has a voltage gain of 1. This means that the Op Amp does not actually provide any amplification to the signal. The reason this exists is that when a circuit such as one with Op Amps has a very high input resistance, very little current is drawn from it. As you know from Ohm's Law, the current is calculated by dividing the voltage by the resistance. Therefore, the greater the resistance, the less current is drawn from a power source. This would therefore affect the power getting drawn.

The reason it is called a voltage-follower is because the output voltage directly follows the input voltage. This allows the output voltage to be the same as the input voltage.

To calculate the output voltage of the circuit, we can use the following equation:

$$V_o = V_i$$

(Voltage out = Voltage in)

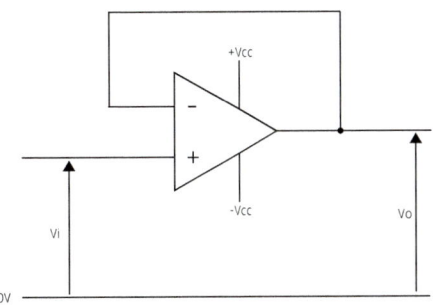

VIDEO LINK

Watch the video at www.brightredbooks.net to learn more about a voltage-follower.

THINGS TO DO AND THINK ABOUT

Calculate the output voltage for the circuit shown.

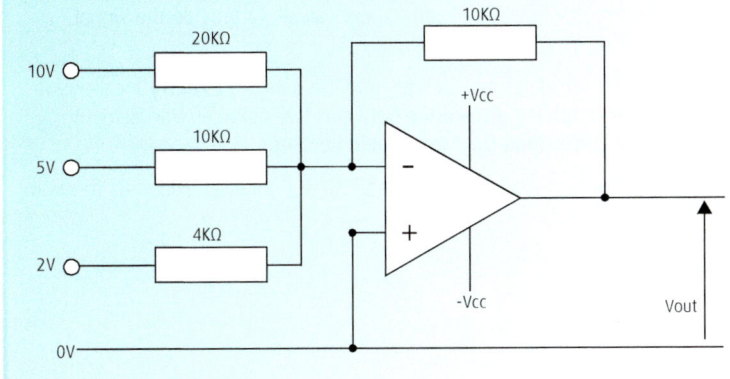

ONLINE TEST

Test your knowledge of this topic at www.brightredbooks.net

ELECTRONICS AND CONTROL

ANALOGUE ELECTRONICS: OPERATIONAL AMPLIFIERS 4

OP AMP CIRCUIT ASSESSMENT

Within any form of assessment throughout this course, you could be asked to do a number of different things with Op Amps. For example, you could be asked to use your mathematical skills to calculate different values within the circuit, you may be asked to design and draw a circuit containing some of the different types available, or you could be asked to write a description of a circuit that contains Op Amps. A number of these could be used to show that your knowledge and understanding of the use of them.

Example 1:

The Diamond Bridge is the third bridge in Aberdeen to cross the River Don, and its construction was necessary to try to manage the high levels of congestion going into and out of the city.

To ensure that it is not being overloaded, the load is monitored using strain gauges on different parts of the bridge. When the bridge is under the maximum permitted strain, the resistance of the active gauge rises to 151 Ω, and the passive gauge remains at 150 Ω. The sensing part of the circuit at maximum permitted strain is shown here.

The measuring instrument requires a voltage of 5 V to give a reading showing the maximum permitted strain. Draw the Op Amp circuit, showing all component values, to provide the signal conditioning described.

To answer this, we first have to look at our data booklet and copy the drawing **exactly** for the type of Op Amp it is. As the question is looking for a **difference** between two voltages (the actual V_{gauge} voltage and the reference voltage), a **difference Op Amp** should be used.

Next, V_{out} can be calculated for both the V_{gauge} voltage-divider and the V_{ref} voltage-divider, as this will give us the inputs into the Op Amp.

V_{gauge}
$V_{out} = (R_2 / R_1 + R_2) \times V_{cc}$
$\phantom{V_{out}} = (151 / 150 + 151) \times 12$
$\phantom{V_{out}} = 6.0199$ V

V_{ref}
$V_{out} = (R_2 / R_1 + R_2) \times V_{cc}$
$\phantom{V_{out}} = (2 / 2 + 2) \times 12$
$\phantom{V_{out}} = 6$ V

These calculations can now be used to work out the gain.
$A_v = V_o / (V_2 - V_1)$
$ = 5 / (6.0199 - 6)$
$ = \underline{\mathbf{251}}$

Now that the exact values of the components are known, these can be added to the drawing. In this case, as the gain is 251, using the calculation $A_v = R_f/R_i$, it can be ensured that the gain will work out like this by using this ratio, 251:1. The simplest way to achieve this is by just making $R_f = 251$ and $R_i = 1$.

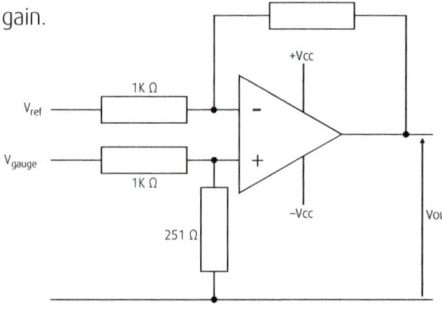

Example 2:

A significant proportion of the marks within the Higher question paper are gained from descriptive answers to questions, so you should ensure that you understand exactly how things work, because one question is likely to be about electronic systems. If you do not fully understand the circuit, break it down into smaller chunks to stop you from becoming confused.

An electrical circuit may include several different Op Amps, and this could be for a multitude of reasons. For example, there may be two different inverting Op Amps to invert the signal back into a positive voltage, while creating an even bigger gain to amplify the signal even more.

In this case, though, it is used for two different outputs. The diagram shows a central heating system that will circulate water at a higher temperature on colder days than it would do on warmer days.

DON'T FORGET

It may be useful when approaching a question like this to answer it in bullet points. This will make it clear and concise not only for the marker but also for yourself.

Electronics and Control: Analogue Electronics: Operational Amplifiers 4

T1 is used to measure the temperature of the water that is circulating within the system, and T2 is used to measure the outside temperature.

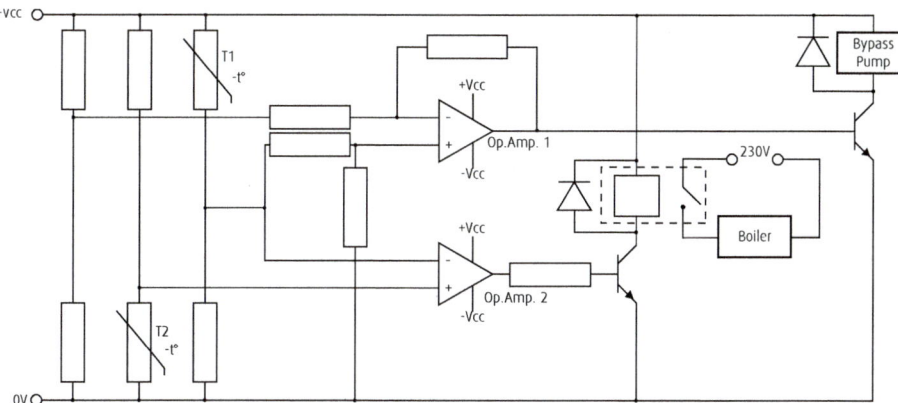

a) Describe how this circuit controls the boiler as the outside temperature decreases.
b) When the boiler switches off, the water within the system may be above a pre-set safe temperature. Describe how this circuit would switch on the bypass pump to circulate the water until it cools down.

a)
- As the outside temperature goes down, the resistance in T2 will go up.
- This means the voltage at the non-inverting input of Op Amp 2 increases.
- The non-inverting input will therefore be greater than the inverting (reference) level.
- This means the Op Amp output will be high.
- This will therefore switch on the transistor.
- In turn, this will switch on the relay, connecting the boiler to the higher-powered circuit, switching the boiler on.

b)
- As the water temperature increases, the resistance in T1 will decrease.
- V_{out} from this voltage-divider then increases, sending a higher signal into Op Amp 1's non-inverting input.
- The first voltage-divider in the circuit (connected to the inverting input of Op Amp 1) acts as the reference voltage.
- When the voltage from T1's voltage-divider becomes higher than the reference voltage going into Op Amp 1, it will cause the Op Amp to saturate and switch on.
- This will then saturate the second transistor, causing it to switch.
- This will then switch on the bypass pump, allowing the water to circulate.
- It will continue to circulate until the temperature cools and V+ in Op Amp 1 is less than V−.

THINGS TO DO AND THINK ABOUT

Go to the SQA past-papers site and attempt the Op Amp questions within the Higher Engineering papers. The answer schemes are also available here, so double-check your answers to ensure you are doing them correctly. If not, study them to see where you are going wrong.

> **DON'T FORGET**
> If you run out of space when answering a question, don't feel pressured to try to squeeze everything into the space given. There will be blank pages at the back of the exam paper for you to use. Just remember to write on the question 'see extra sheet' or something similar, so that the marker knows to look for your complete answer.

> **ONLINE**
> Head to www.brightredbooks.net for a link to the Higher Engineering SQA Past Papers.

> **ONLINE TEST**
> Test your knowledge of this topic at www.brightredbooks.net

ELECTRONICS AND CONTROL

ANALOGUE ELECTRONICS: TRANSISTORS

VIDEO LINK

Go to www.brightredbooks.net to watch a video that will help you obtain a greater understanding of how a transistor works.

AMPLIFICATION

Input transducers that produce a voltage rarely produce sufficient voltage for most applications. This means that their outputs have to be amplified. One of the most commonly used amplifying devices in an electronic system is the **bipolar junction transistor** (BJT). There are two types of BJT transistors that are available for use, and these are the PNP transistor and the NPN transistor.

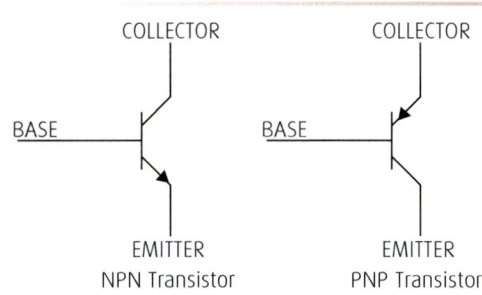

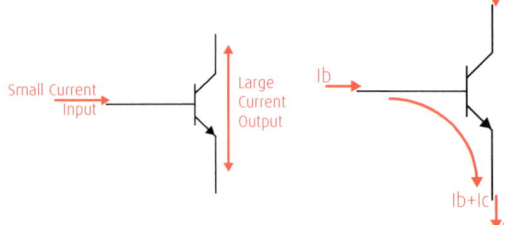

The difference between them is that an NPN transistor will amplify a positive signal applied to the base by allowing a larger current to flow from the collector to emitter, proportional to the base voltage. In a PNP transistor, negative voltage must be applied to the base to allow current to flow from the emitter to the collector.

As you have previously learned in National 5, a small current flowing between the base and emitter of a transistor allows a large current to flow between the collector and emitter. Knowing this, it can be deduced that **Ie = Ib + Ic.**

Since I_b is usually much smaller than I_c, it means that I_e must be approximately the same value as I_c

$$I_e = I_c$$

The current gain (or amplification) of the transistor is defined as the ratio of collector current to base current. That means we can calculate the gain the transistor has by dividing the collector current by the base current:

$$h_{FE} = I_c / I_b$$

(Current gain = Collector current ÷ Base current)

For example, if a current of 10mA is measured at the collector of a transistor when the base current is 0.25mA, the gain is:

$$h_{FE} = I_c / I_b$$
$$= 10 / 0.25 = \underline{\mathbf{40}}$$

DON'T FORGET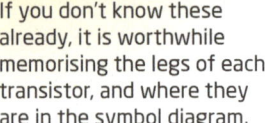

If you don't know these already, it is worthwhile memorising the legs of each transistor, and where they are in the symbol diagram.

TRANSISTORS IN CIRCUITS

In order to generate a current into the base of the transistor, a voltage must be applied at the base-emitter junction (V_{be}). No current will flow if V_{be} is below 0.7V as the transistor will be off. When the input to the base reaches 0.7V, the transistor will be saturated (switched on) and will then act as a current amplifier. Any further increase to the base will not increase the V_{be} value, and it will not increase the collector current.

Any question that contains a transistor, you should assume that V_{be} is 0.7V whenever the transistor is on.

Example 1: transistor amplifier circuit

You may be asked to calculate unknown quantities in a transistor circuit like the one shown. In this case, you are asked to calculate the light level sensed by the LDR to produce a current of 30mA in the 5V bulb. The HFE of the transistor is 20.

contd

First, calculate the base current of the transistor:

$h_{FE} = I_c / I_b$
$\Rightarrow I_b = I_c / H_{FE}$
$= 30 \text{ mA} / 20$
$= \mathbf{1 \cdot 5 \text{ mA}}$

Now calculate the voltage (V_{out}) at the centre of the voltage divider. The transistor is on, so we can assume that V_{be} is 0.7V. We know the base current so can calculate the voltage across the 680Ω base resistor using Ohm's Law:

V = IR
$= (1 \cdot 5 \times 10 - 3) \times 680 = \mathbf{1 \cdot 02 \text{ V}}$
$V_{out} = 0 \cdot 7 \text{ V} + 1 \cdot 02 \text{ V} = 1 \cdot 72 \text{ V}$

We can then calculate the current through the 20 kΩ resistor.

$V = IR \Rightarrow I = V / R$
$= 1 \cdot 72 / 20 \times 10^3 \text{ A} = 0 \cdot 000086 \text{ A} = 0 \cdot 09 \text{ mA}$
The current through the LDR,
$I_{LDR} = 1 \cdot 5 \text{ mA} + 0 \cdot 09 \text{ mA} = 1 \cdot 59 \text{ mA}$

The voltage across the LDR
$V_{LDR} = 9 \text{ V} - 1 \cdot 72 \text{ V} = 7 \cdot 28 \text{ V}$

The resitance of the LDR = V / I

$= 7 \cdot 28 / (1 \cdot 59 \times 10 \cdot 3) = 4 \cdot 6 \text{ k}\Omega$

Knowing this, the LDR graph can be used to discover that the light level is **20 lux approx**.

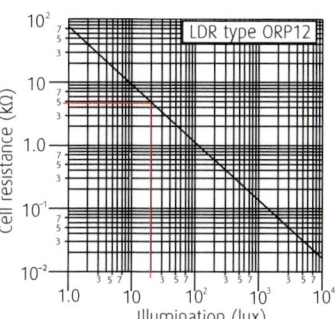

Example 2: Transistor switching circuit

As part of a city-wide regeneration project, Aberdeen is planning on installing a water feature in a new civic square. As well as being a feature attraction, in the warmer weather it can act as a childrens play area so that when a child blocks the light to a darkness sensor, it will spray.

The circuit diagram is shown opposite.

a) Calculate the current flowing through the LDR.
b) Calculate the current flowing through the variable resistor.
c) Calculate the base current flowing into the transistor.

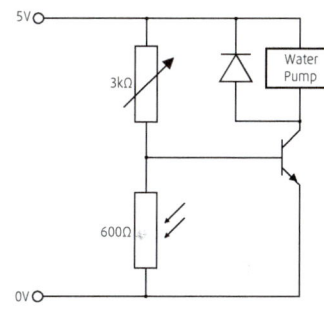

a) As we know, there must be 0.7V across the base emitter junction to switch the transistor on. This is the same as the voltage across the LDR. The LDR current can then be calculated using Ohm's Law:
V = IR $\Rightarrow$ I = V / R
$= 0 \cdot 7 / 600 = \mathbf{0 \cdot 00116 \text{ A}} = \mathbf{1 \cdot 2 \text{ mA}}$ (S.F.)
b) The supply voltage is 5 V, and the LDR has 0.7 V across it. Knowing this we can calculate the volts across the variable resistor:
V = 5 V − 0·7 V = **4·3 V**

The current through the variable resistor can then be calculated using Ohm's Law:

V = IR $\Rightarrow$ I = V / R
$= 4 \cdot 3 / 3,000 = \mathbf{0 \cdot 00143 \text{ A}} = \mathbf{1 \cdot 4 \text{ mA}}$ (S.F.)
c) Ib = 1·4 mA − 1·2 mA = **0·2 mA**

 THINGS TO DO AND THINK ABOUT

The circuit shown here is being designed by an electronics engineer to switch on a heated seed-propagator when the temperature falls to 14°C.

At 14°C, with the transistor saturated, the resistance of the type 2 thermistor is 6 kΩ, and the current through the 150 kΩ resistor is 0·25 mA.

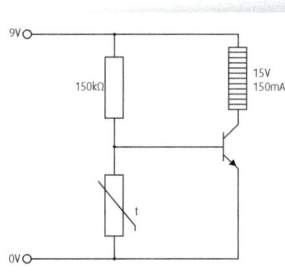

a) Calculate the current flowing through the thermistor.

b) Calculate the current flowing through the base of the transistor.

c) Calculate the minimum current gain required for heating the element to fully switch on at 14°C.

ONLINE TEST

Test your knowledge of this topic at www.brightredbooks.net

ELECTRONICS AND CONTROL
ANALOGUE ELECTRONICS: MOSFETS

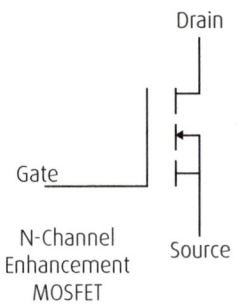

WHAT IS A MOSFET?

A way of driving larger loads is the inclusion of a MOSFET. In National 5, you learned that if a larger load was needed, it would have to be connected to a higher power supply via a relay, but in the real world this is not always practical or suitable - instead a MOSFET can be used. A MOSFET works in the same way as a transistor, but unlike a bipolar junction transistor, which is a current-operated device, a MOSFET is voltage-operated. MOSFETs also have a very high input resistance, unlike a BJT, which means that a MOSFET only needs a very small current for it to work.

Like a BJT, a MOSFET will switch when a certain voltage is reached, but this threshold voltage varies depending on which MOSFET you are using. It **<u>does not</u>** switch at 0·7 V like a normal bipolar junction transistor!

Instead of its legs being known as the base, collector and emitter, a MOSFET's legs are known as the **gate**, **drain** and **source**. By applying a voltage at the gate of the MOSFET, it generates an electrical field to control the current flow through the channel between drain and source, and there is no current flow from the gate into the MOSFET.

N-Channel Enhancement MOSFET

Like a transistor, a MOSFET also comes in two different types: an N-type and a P-type, with the N-type being the most commonly used, and the one you will learn about in this course. In an N-channel MOSFET, the field channel is negative. This means that, like a transistor, the bottom leg (the source) is connected to the ground, and when the input (the gate) is connected to this, it is 'off'. This essentially means that when the gate voltage is low, the circuit is incomplete.

To turn the MOSFET on and put it into 'enhancement mode', the voltage on this gate needs to be raised, making the gate voltage (V_G) higher than the source voltage (V_S). When the gate voltage reaches the threshold voltage, a channel can be produced, and the circuit completes.

VIDEO LINK

Watch the video on the BrightRED Digital Zone to obtain a greater understanding of how a MOSFET works.

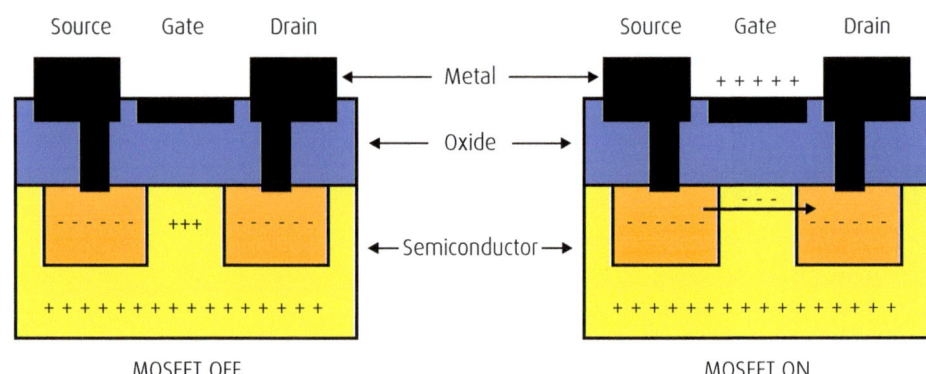

MOSFET OFF   MOSFET ON

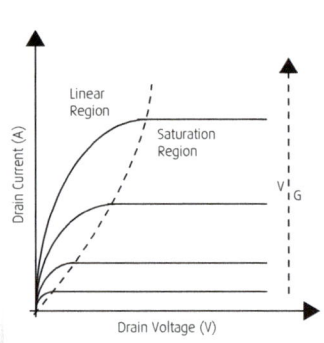

When the MOSFET is in the linear region/mode, the device will conduct better, as the gate voltage is greater.

MOSFET Saturation Point

As you have already learned, for any given MOSFET, the size of the current between the drain and source will therefore depend on the gate voltage (V_{GS}) and the voltage between the drain and source (V_{DS}).

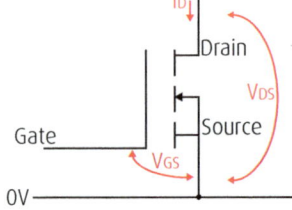

Like a BJT, if the input voltage (V_G) is below a certain level (the threshold value, V_T), it will not allow a channel to open. This means the current will not flow and the MOSFET will switch off. If the gate voltage is above this V_T, the MOSFET will switch on.

This means that, for any given value of V_{GS}, increasing V_{DS} will increase the current until saturation occurs. Any further increase will not increase I_D.

ONLINE

Visit www.brightredbooks.net and find a link that will help you obtain a greater understanding of N-channel MOSFET characteristic curves.

contd

Knowing this gives you the calculation:

$V_{DS} = V_{GS} - V_T$

When the saturation point is reached, the current at I_D is the same as it was before saturation. This means that:

$I_D = I_{D(ON)}$

Example:

The gate-voltage threshold for the MOSFET shown is 3 V. Calculate the gate voltage required to ensure that a saturation current of 10 mA flows through the load resistor.

The drain–source channel essentially acts like a resistor in series with the 150 Ω resistor – and, since the current in a series circuit is the same throughout, the voltage can be calculated using Ohm's Law:
V = IR = 10 mA × 150
= **1·5 V**

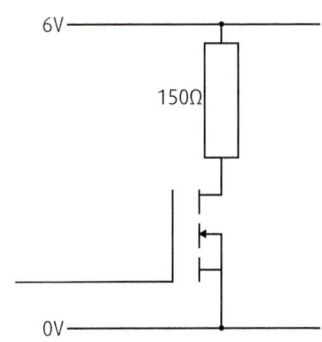

To calculate the voltage over the channel, we can subtract the voltage going over the resistor from the supply voltage:
V_{DS} = 6 − 1·5
= **4·5 V**

Now we can use our MOSFET saturation calculation to work out the gate voltage required for this MOSFET to saturate:
$V_{DS} = V_{GS} - V_T$ => $V_{GS} = V_{DS} + V_T$ = 4·5 + 3 = **7·5 V**

MOSFETs Used as a Driver

Unlike BJTs, MOSFETs make very good drivers, as they can handle very high drain currents. This means that they can be used to drive high-current-output transducers without the need for a relay or adding extra components.

A variable resistor is used here in a voltage-divider circuit. This allows the resistance to be adjusted to ensure that the input voltage to the gate = V_T.

Changes in V_{GS} (ΔV_{GS}) that are above the threshold value will cause a change in I_D (ΔI_D). This is known as the **transconductance** (g_m) and can be calculated by:

$g_m = \Delta I_D / \Delta V_{GS}$

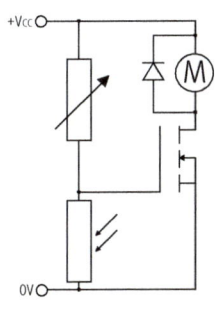

> **DON'T FORGET**
>
> MOSFETs are particularly sensitive to high voltages, so be careful always to include a diode over any transducers that may cause back-EMF when switched off.

Example:

Calculate I_D and V_{GS}, and hence the transconductance of the MOSFET in the configuration shown.
To work out the drain current, Ohm's Law can be used:

V = IR => I = V/R = 7 / 3,300 I_D = **2·12 mA**

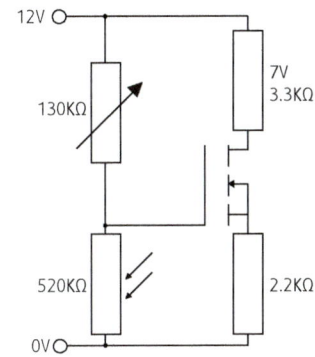

To work out the V_{GS}, you first have to calculate the source voltage and the gate voltage.
V_S can be calculated also using Ohm's Law:
$V_S = I_D \times R_S$ = 2·12 × 2·2 = **4·67 V**

The gate voltage can be calculated using the voltage-divider V_{out} calculation:
$V_G = R_2 / (R_1 + R_2) \times V_{CC}$ = (520 / (130 + 650)) × 12 = **9·6 V**

$V_{GS} = V_G - V_S$ = 9·6 − 4·67 = **4·93 V**

$g_m = \Delta I_D / \Delta V_{GS}$ = 2·12 / 4·93 = **0·43 AV^{-1}**

THINGS TO DO AND THINK ABOUT

1. An electronics engineer is designing the control system for a juicer.

    a) In what ways do MOSFETs and bipolar junction transistors differ in controlling their output current?

    b) A MOSFET was chosen to drive the electric motor. Why would this be chosen instead of a bipolar junction transistor?

ELECTRONICS AND CONTROL

DIGITAL ELECTRONICS: LOGIC GATES AND BOOLEAN EXPRESSIONS

ONLINE

Improve your knowledge of logic-gate symbols by visiting www.brightredbooks.net

DON'T FORGET

It is worth memorising these gate symbols and making sure you fully understand exactly how they work. This is the basis for all digital electronics in the Higher course.

VIDEO LINK

Go to www.brightredbooks.net and watch a video to remind yourself of how these three gates work.

LOGIC GATES

Although it may not always seem like it, electronic systems are very logical in the way that they work. Essentially, if you want a light to come on, you press a switch, and the light should come on – but, as you already know, it's actually a lot more complex than that. Most systems involve making much more complicated decisions, such as sorting out bottles in a recycling plant, checking the room temperature in a central heating system, or switching on a security alarm if movement is sensed when there shouldn't be any.

Logic gates are frequently used in systems like this, as they are components used for processing a combination of different inputs. In National 5, you have already gained experience in the use of **NOT**, **AND** and **OR** logic gates.

NOT Gate

In essence, a NOT gate inverts a signal – because of this, it is sometimes known as an **inverter**. If the input to the gate is high, the output is **not** high (i.e., it is low); and the same can be said if the opposite happens.

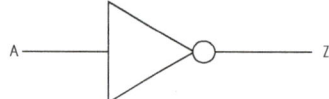

A	Z
0	1
1	0

AND Gate

An AND gate will only give a high output signal if input A **and** input B are high.

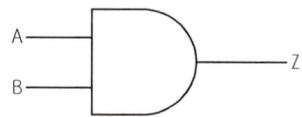

A	B	Z
0	0	0
0	1	0
1	0	0
1	1	1

OR Gate

An OR gate will only give a high output signal if input A **or** input B is high.

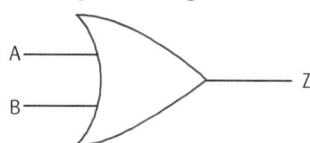

A	B	Z
0	0	0
0	1	1
1	0	1
1	1	1

Within the Higher course, there are three other gates that you will **need** to learn: the **NAND**, **NOR** and **EOR** gates.

NAND Gate

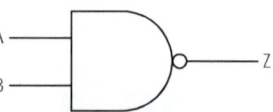

This should be seen as a '**NOT** AND' gate, as the outcomes are the opposite of what you would expect with a normal AND gate.

As you can see, the symbol is a combination of a NOT gate and an AND gate symbol. It has the body of the AND gate but the circle of a NOT gate to show that the signal is inverted.

A	B	AND gate	Z (NAND)
0	0	0	1
0	1	0	1
1	0	0	1
1	1	1	0

contd

NOR Gate

This should be seen as a '**NOT** OR' gate, as the outcomes are the opposite of what you would expect with an OR gate.

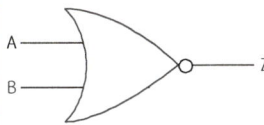

As you can see, the symbol is a combination of NOT gate and OR gate symbols. It has the body of the OR gate, but the circle of a NOT gate to show that the output signal will be inverted.

A	B	OR gate	Z (NOR)
0	0	0	1
0	1	1	0
1	0	1	0
1	1	1	0

EOR Gate

The EOR gate (sometimes known as an XOR gate) should be seen as an '**EXCLUSIVE** OR'. This is because a high output is only given when it is Exclusively one input on, **OR** the other. An OR gate will give a high output when both inputs are high, but an EOR gate will not do this. With an EOR, it must be **E**ither one **OR** the other.

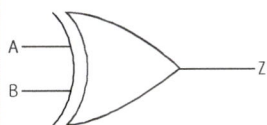

As you can see, the symbol is very similar to an OR gate symbol, but it has an extra line. Think of this as if it is being double-underlined, to remind you that it will only work Exclusively with one input on **OR** the other.

A	B	OR gate	Z (EOR)
0	0	0	0
0	1	1	1
1	0	1	1
1	1	1	0

BOOLEAN EXPRESSIONS

Each logic gate has a corresponding mathematical equation. This is known as the **Boolean expression**. This Boolean expression allows us to know how a circuit works by a simple equation, rather than having to draw out a whole circuit.

NOT GATE: $Z = \overline{A}$

AND GATE: $Z = A.B$

OR GATE: $Z = A+B$

NAND GATE: $Z = \overline{A.B}$

NOR GATE: $Z = \overline{A+B}$

EOR GATE: $Z = A \oplus B$

THINGS TO DO AND THINK ABOUT

Create the game of 'Snap' with cards that contain the logic symbol, as well as other cards that contains the Boolean expression. Play with your classmates and see who wins!

DON'T FORGET

To help you memorise these, look at what's different. If it has a circle on the end like a NOT gate, it is inverting the signal. If it has the extra line like the EOR gate, see this as double underlining it to emphasise that it must be one **or** the other.

ONLINE

Improve your knowledge of logic gates by visiting the link on the BrightRED Digital Zone.

DON'T FORGET

It is worth memorising these Boolean expressions, as it should be expected that these will appear in any examination.

DON'T FORGET

Don't fall into the trap of using + for ANDs, as this will give you the wrong answer. Remember: AND is . and OR is +.

ONLINE TEST

Test your knowledge of this topic at www.brightredbooks.net

ELECTRONICS AND CONTROL

DIGITAL ELECTRONICS: COMBINATIONAL BOOLEAN

DISCOVERING BOOLEAN FROM A CIRCUIT

A digital electronic circuit is likely to utilise a combination of different types of logic gates. This is known as **combinational logic**. Boolean expressions can then be worked out from these to find the overall equation for the circuit.

Example:

In a circuit such as this, an overall equation can easily be calculated by knowing the corresponding Boolean equation for each logic gate. To work this out, it needs to be taken one step at a time, as this will allow you to work out the partial equation as it goes through each logic gate. It is also advisable that you write on each path what the new name of it is. This will stop you from being confused when it starts to become more complicated.

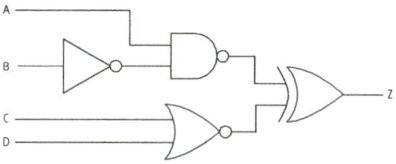

Step 1: Start from the beginning of the circuit, and see what the most obvious gate is to do first. In this case, the path of A is connected to a NAND gate, but B goes into another logic gate before it reaches this. That means the output of B has to be determined before the output of this NAND gate can be calculated. This means that the obvious first step should be determining the outcome of B.

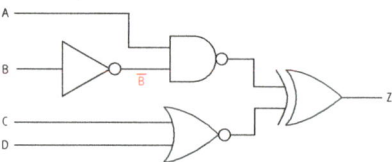

Step 2: As we now see, the signals going into this NAND gate are A and $\bar{B}$. Knowing this, we can now determine the name of the path leaving the NAND gate.

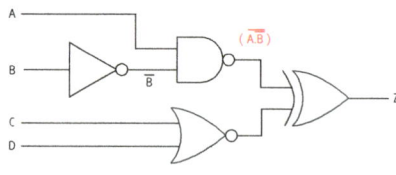

Step 3: Moving down, we can see there is a NOR gate into which the signals C and D are fed – and we can now discover what the output of the gate is using them.

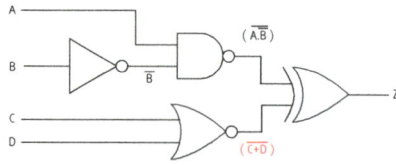

Step 4: Finally, we know that the two inputs to the EOR gate are $\overline{(A.\bar{B})}$ and $\overline{(C + D)}$, so we can now solve the output to this gate and, in turn, decipher the Boolean expression for the whole circuit.

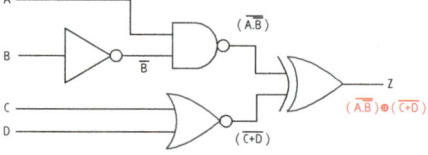

This means that the final Boolean expression is: $Z = \overline{(A.\bar{B})} \oplus \overline{(C+D)}$

DON'T FORGET

It is a good idea to put brackets around the outcomes when it contains more than one signal. This is to help show that this part should now be seen as one whole signal, and not two separate quantities.

VIDEO LINK

To obtain a greater understanding of this, watch the video on the BrightRED Digital Zone.

DISCOVERING BOOLEAN FROM A BRIEF

When asked to work out the Boolean expression, sometimes the circuit is not given. Instead, a brief could be provided letting you know what the circuit is supposed to do.

Example:

A person improves their own home security by installing a security camera at their front door. The camera is activated when the door is opened, or when it is dark and someone is sensed near the door. Write a Boolean expression in terms of the three inputs.

Door sensor – INPUT A (A = 1 when door is closed)

Light sensor – INPUT B (B = 1 when it is light outside)

Person sensor – INPUT C (C = 1 when person is sensed)

Camera – OUTPUT Z (Z = 1 when camera is activated)

To approach a question such as this, you have to make sure you are reading the brief properly. It states that 'The camera is activated when the door is opened'. So, if A is high when the door is closed, this must be NOT A.

$Z = \overline{A}$

After this, it uses the word 'or'. This means we must be using an OR gate next.

$Z = \overline{A} +$

The next section states that it will work 'when it is dark'. As B = 1 when it is light outside, it must work when it is NOT B, so B must be inverted. After this, it says: 'and someone is sensed near the door'. As the word 'and' is used, it must be assumed an AND gate will be needed to join the two parts of this section together. The second part will be C, as it is referring to someone being sensed, and when someone is sensed C will be high. This gives us the final Boolean expression of:

$Z = \overline{A} + (\overline{B}.C)$

> **DON'T FORGET**
>
> If you find it useful to do so, underline or use a highlighter to highlight important parts of the brief to make it clearer for you.

THINGS TO DO AND THINK ABOUT

Calculate the Boolean expression for the following circuits:

a)

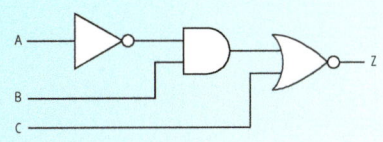

b)

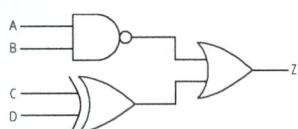

c)

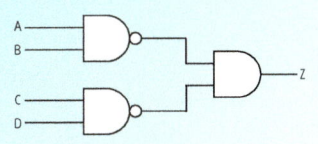

> **ONLINE TEST**
>
> Test your knowledge of this topic at www.brightredbooks.net

ELECTRONICS AND CONTROL

DIGITAL ELECTRONICS: COMBINATIONAL LOGIC TABLES

DON'T FORGET

Take your time filling out truth tables as it is easy to make a mistake and get confused when dealing with only 1's and 0s's. Fold your paper over or use a ruler to block out unneeded columns if it helps you focus.

WORKING OUT THE TRUTH TABLE

For each combinational logic diagram, a truth table can also be determined. This can look confusing at first, but it can be very simple if time is taken and it is broken down into smaller chunks. The first thing to do is to name each path that is an output from any logic gate, and then to create a truth table containing not only the input and output paths from the complete circuit, but also those that are the outputs from each individual gate.

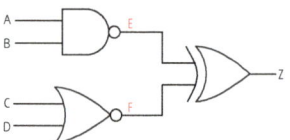

A	B	C	D	E	F	Z

Now the table can be partially completed by putting in **every single** possible combination for the four inputs. The easiest way to work this out is to use the two possible combinations (0 and 1) and put this to the power of however many inputs to the circuit there are. In this circuit, there are four possible inputs (A, B, C and D). So, by working out 2^4, we can discover that there should be 16 combinations, and therefore 16 different rows.

As a normal two-input logic gate can have four possible input conditions (0-0, 0-1, 1-0, 1-1), each option should be repeated four times (4 × 4), and that would give the needed 16 rows. We can use this to help work out what all the possible combinations are:

A	B	C	D	E	F	Z
0	0	0	0			
0	0	0	1			
0	0	1	0			
0	0	1	1			
0	1	0	0			
0	1	0	1			
0	1	1	0			
0	1	1	1			
1	0	0	0			
1	0	0	1			
1	0	1	0			
1	0	1	1			
1	1	0	0			
1	1	0	1			
1	1	1	0			
1	1	1	1			

Now the solution has to be worked out in different stages. As you can see from the diagram, E is the output of A and B through a NAND gate, so that is the next column to resolve.

contd

Electronics and Control: Digital Electronics: Combinational Logic Tables

In this example, the unwanted columns have been blacked out to help you in the understanding.

A	B	E
0	0	1
0	0	1
0	0	1
0	0	1
0	1	1
0	1	1
0	1	1
0	1	1
1	0	1
1	0	1
1	0	1
1	0	1
1	1	0
1	1	0
1	1	0
1	1	0

Column F can also be deciphered, as this is the result of C and D going through a NOR gate.

C	D	F
0	0	1
0	1	0
1	0	0
1	1	0
0	0	1
0	1	0
1	0	0
1	1	0
0	0	1
0	1	0
1	0	0
1	1	0
0	0	1
0	1	0
1	0	0
1	1	0

And finally, the output column Z can be discovered, as this is the outcome of E and F going into an EOR gate.

E	F	Z
1	1	0
1	0	1
1	0	1
1	0	1
1	1	0
1	0	1
1	0	1
1	0	1

E	F	Z
1	1	0
1	0	1
1	0	1
1	0	1
0	1	1
0	0	0
0	0	0
0	0	0

VIDEO LINK

Watch the video at www.brightredbooks.net to obain a better understanding of how to create truth tables.

THINGS TO DO AND THINK ABOUT

Determine the Boolean expressions for the following circuits, and work out the truth tables. Take these into class, and get your classmates to check if you are correct.

a)
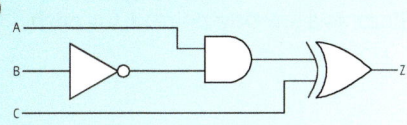

b)
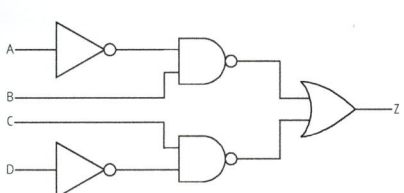

c)
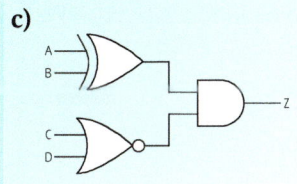

ONLINE TEST

Test yourself on this topic at www.brightredbooks.net

ELECTRONICS AND CONTROL

DIGITAL ELECTRONICS: CREATING BOOLEAN FROM TRUTH TABLES

COMBINATIONAL LOGIC TABLES

When designing systems, it is common practice for an engineer to design a logic diagram from a prepared truth table. This may seem difficult to start with, but if you concentrate on the combinations which only give a logic 1 condition in the output column, solutions can be found easily.

DON'T FORGET

If it is easier, use a highlighter pen to colour the rows that will switch the input on. This will help you focus on the important aspects of the truth table, and stop confusion.

This truth table shows three inputs, A, B and C, and one output, Z.

A	B	C	Z
0	0	0	0
0	0	1	0
0	1	0	0
0	1	1	1
1	0	0	0
1	0	1	0
1	1	0	1
1	1	1	0

Output Z is at logic 1 in two places – firstly in the fourth row down, and we can see that for this to happen A must be at logic 0, B must be at logic 1, and C must be at logic 1. In other words:

Z = NOT A **and** B **and** C
Z = $\overline{A}.B.C$

Output Z is also at logic 1 at the second-bottom row. For this condition to be true, A must be at logic 1, B must be at logic 1, but C must be at logic 0. In other words:

Z = A **and** B **and** NOT C
Z = $A.B.\overline{C}$

If the first condition **or** the second condition is true, then Z switches on. So...

Z = $(\overline{A}.B.C) + (A.B.\overline{C})$

VIDEO LINK

Head to www.brightred.books.net and watch the video to get a greater understanding on how to create logic circuits.

CREATING CIRCUIT DIAGRAMS FROM BOOLEAN EXPRESSIONS

If we know the Boolean expression, it is then possible to draw the circuit diagram. To draw the circuit, it is always a good idea to draw vertical lines as a starting point: 1 for each circuit input that exists.

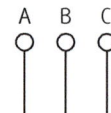

Now we break down the Boolean expression. The first part of the equation is $(\overline{A}.B.C)$. We know that it is NOT A, so we can connect a NOT gate to this line.

contd

Electronics and Control: Digital Electronics: Creating Boolean from Truth Tables

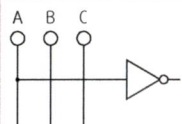

We also know that the three inputs get ANDed together. To do this, we can use a 3-input AND gate and draw this.

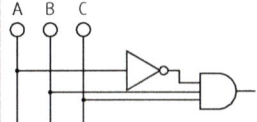

We can now repeat this for the second half of the equation (A.B.$\overline{C}$) in the exact same fashion. This should be separate from the first diagram, but still connecting to the same wires.

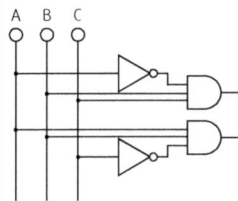

As we know, either the top circuit **or** the bottom circuit will cause the outcome conditions to be true, so we can connect these using an OR gate.

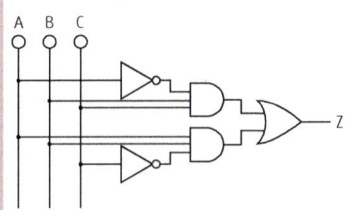

THINGS TO DO AND THINK ABOUT

For the following logic tables, work out the Boolean expression and draw its corresponding circuit diagram.

ONLINE TEST

Test your knowledge of this topic at www.brightredbooks.net

a)

A	B	C	Z
0	0	0	0
0	0	1	0
0	1	0	0
0	1	1	1
1	0	0	0
1	0	1	0
1	1	0	0
1	1	1	1

b)

A	B	C	D	Z
0	0	0	1	0
0	0	1	0	1
0	1	0	1	0
0	1	1	0	1
1	0	0	1	1
1	0	1	0	0
1	1	0	1	1
1	1	1	0	0

c)

A	B	C	D	Z
0	0	0	1	0
0	0	1	1	1
0	1	0	1	1
0	1	1	0	0
1	0	0	1	0
1	0	1	0	1
1	1	0	1	1
1	1	1	0	0

ELECTRONICS AND CONTROL

DIGITAL ELECTRONICS: NAND EQUIVALENTS

NAND GATES

NAND gates are called **functionally complete**. This means that they are able to be used to express all possible truth tables. Because of this, combinations of NAND gates can be used to replicate the function of any other gate. Using these as a universal gate makes it extremely beneficial to industry, as buying a couple of NAND chips would be cheaper than buying four or five different ICs of different gates. This also means that it would take up a smaller amount of space in an electronic system.

DON'T FORGET

It is highly advisable that you memorise these configurations, as it is very likely you would be expected not only to know these for any assessment but also to be able to convert basic logic systems into NAND-only systems.

NOT Gate — NOT Gate NAND Equivalent

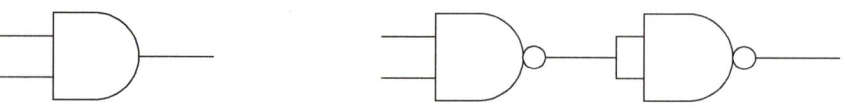

AND Gate — AND Gate NAND Equivalent

VIDEO LINK

Head to the Digital Zone to watch a video on how the NAND gate can be used.

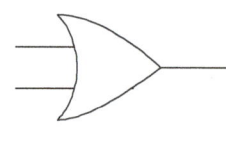 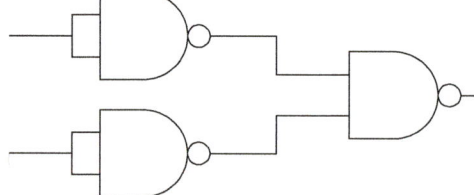

OR Gate — OR Gate NAND Equivalent

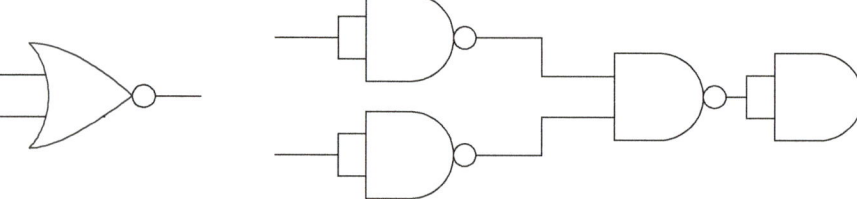

NOR Gate — NOR Gate NAND Equivalent

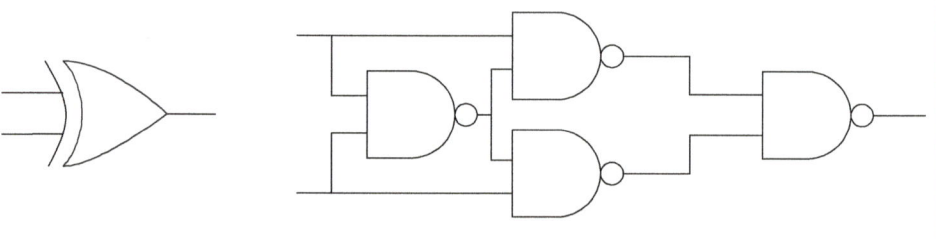

EOR Gate — EOR Gate NAND Equivalent

CONVERSION TO NAND EQUIVALENTS

In the question paper, it is highly likely you will have to change a logic system to its NAND equivalent – but, if you have memorised these configurations, it shouldn't be difficult. Consider the circuit shown.

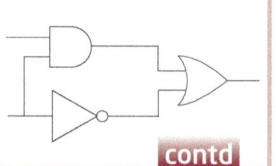

contd

38

Electronics and Control: Digital Electronics: NAND Equivalents

This logic system is made from a NOT gate, an AND gate and an OR gate. To convert this system, firstly redraw the circuit, replacing each gate with its NAND equivalent.

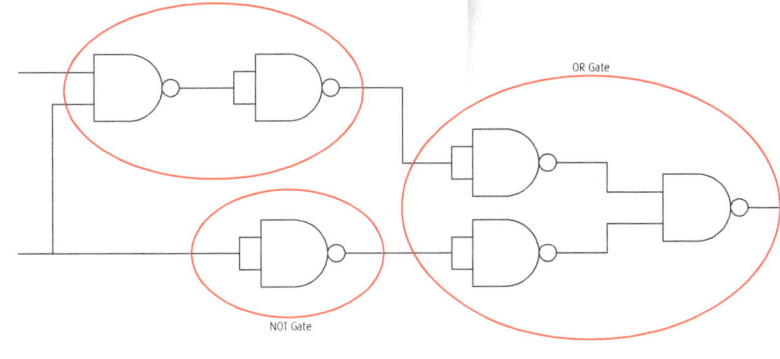

Examine the new arrangement. The next step is to look for adjacent pairs of NOT equivalent gates that have one input split into two, followed by another NOT gate that has one input split into two.

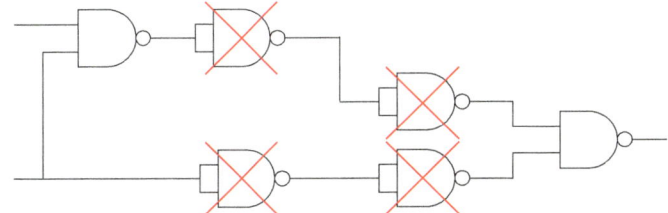

If you consider what happens when you feed a signal to a NOT gate and then pass the signal on to another NOT gate, you will find that the signal has been **double inverted**. This means the gates would cancel each other out and therefore would not be needed in the circuit.

In this example, there are two such pairs – these should be scored out.

🤔 THINGS TO DO AND THINK ABOUT

A tumble dryer in a laundromat gets rid of excess heat by venting warm air back into the room. It also collects the moisture from the clothes in a water tank, and an airflow sensor detects when the filter needs to be cleaned.

Water tank – input A
Airflow control – input B
Start button – input C
Buzzer – output Z

A logic system controls a warning buzzer. If the water tank is full (A = 1) or the airflow is too low (B = 0), a buzzer will sound (Z = 1) when the start button is pressed (C = 1).

a) Draw a truth table for the output Z in terms of the inputs A, B and C.

b) Write a Boolean expression for Z in terms of A, B and C.

c) Draw a logic diagram for the control system, constructed from AND, OR and NOT gates.

d) Draw a logic diagram for the control system, using only NAND gates. Simplify where appropriate.

➕ DON'T FORGET

Do not redraw the circuit without the unneeded NAND gates, or completely block them out. It is important that they are just scored or crossed out, as removing them is likely to be worth a mark in the exam. If the marker cannot see that you have removed them, and exactly what you have removed, you will not gain this mark!

✓ ONLINE TEST

Test your knowledge of this topic at www.brightredbooks.net

ELECTRONICS AND CONTROL
PROGRAMMABLE CONTROL: FLOWCHARTING

When creating a complex piece of microcontroller code, flowcharts are commonly used to show how it works. This allows an engineer to work out **exactly** what the program is going to do, and how it is going to work. As flowcharts are drawn graphically, they often make programs easier to understand.

SYMBOLS

Start/Stop Symbol

The 'start/stop' symbol is a sausage-shaped object. Each flowchart **must** contain one start symbol at the beginning – and, if the program is to stop after it has completed a cycle, it will have a stop symbol at the end. If it is to continually loop (like most modern systems), it will not contain a stop symbol. Instead, it will have a continuous loop arrow linking it back to the beginning.

Wait Symbol

The 'wait' symbol is a rectangle. The text inside the symbol should explain how long the time delay is. The unit **must** be mentioned here, or you will lose marks in the exam.

Output Symbol

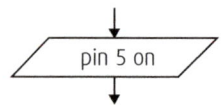

The 'output' symbol is a parallelogram. The text inside the symbol should explain which output pins are switched on or off at any time. The pin number should be mentioned here.

Decision Box

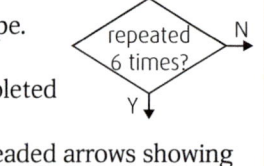

The decision box is a diamond shape. The program uses this to check whether something has been completed or whether something needed has happened. This box should have headed arrows showing the direction of the two different options, and the options **must** be labelled either 'Yes' and 'No' or 'Y' and 'N' to show what each path is for.

Sub-procedure

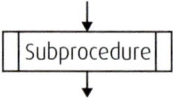

A sub-procedure is a small section of code that can be called upon, but is not part of the main program. Because of this, it should have a separate name that can be referred to in the code. After the sub-procedure is finished, it should then have the same symbol at the bottom of the flowchart, with 'return' written in it. This will then link it back to the main program and allow the code to continue.

DON'T FORGET

Make sure you ALWAYS refer to the pins if it asks for it in the question, and not the actual input or output device. If you refer to the devices in an assessment, and it has asked for the pins, you will not get the marks, even if it is correct!

Input connection	Pin	Output connection
	2	Red light
	1	Amber light
	0	Green light

CONTINUOUS LOOPS

Quite frequently a program is designed never to stop, and will continually run. To do this, it is necessary to have an infinite loop. An example of the use of infinite loops would be traffic lights – you would not want the program controlling them to end as soon as it has gone through each light once. Instead, you want it to loop back to the beginning of the program, and then repeat forever.

Example:

A set of temporary traffic lights is used to control the flow of traffic in roadworks. The operation of the system is as follows:

- The red light goes on for 10 seconds.
- The amber light then comes on for 4 seconds.
- The red and amber lights go off, and the green light switches on for 15 seconds.
- The green light switches off, and the amber light goes on for 4 seconds.
- The amber light goes off, then the system repeats.

Referring to all pins, draw a flowchart to show the control of the traffic lights.

To answer a question like this, go through the specification one by one, making sure it is doing everything asked, the correct symbols have been used, and all the pins have been referred to.

The flowchart should look like this:

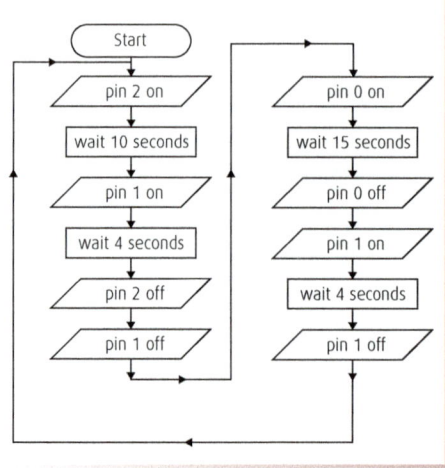

Electronics and Control: Programmable Control: Flowcharting

SUB-PROCEDURES

Within a complex program, sub-procedures are also commonly used to break it up and simplify it. These sub-programs can then be called upon and used to perform specific tasks.

For example, a shop uses a microcontroller-operated goods elevator to transport its products between three floors – the basement where its storage is, and the first and top floors of the building. The flowcharts here show the control sequence for the movement of the lift to the appropriate floor.

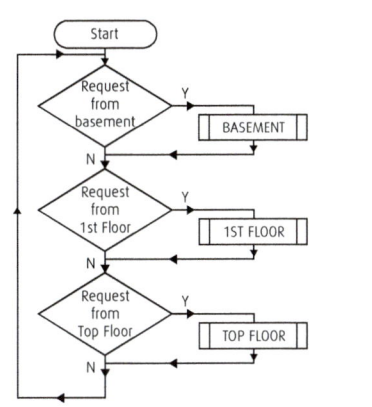

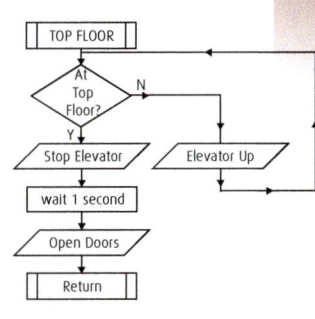

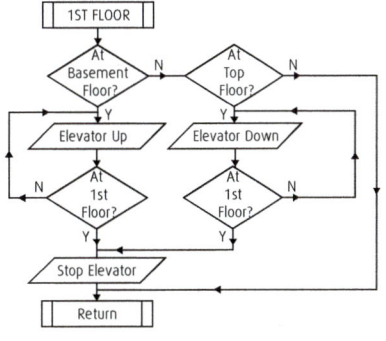

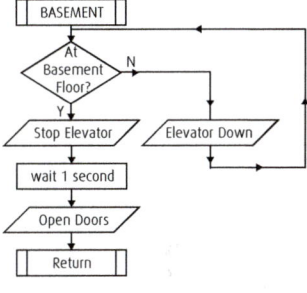

Within any assessment, it is highly likely that flowcharts will appear, but what you have to do with them could vary. You could either be asked to write a full or partial flowchart, simulate one, or it could be a case of having to read one and to answer questions to show your understanding of it.

For example, under what circumstances will the elevator move downwards? By looking through the flowcharts, you should be able to see that it will move under two conditions:

1. if a request is received from the basement, and the elevator is not already at the basement
2. if a request is received from the 1st floor, and the elevator is on the top floor.

THINGS TO DO AND THINK ABOUT

After components have been assembled in a factory, an automated system is used to quality-assure the items.

Part of the system operates on the following sequence:

- The motor (pin 3) must switch on to move the items along a conveyor belt
- A light sensor (pin 0) detects when they have reached the inspection point (a high signal indicates that the door is open)
- A pneumatic piston (pin 4) out-strokes for 1·5 seconds and diverts every fourth item for inspection
- If an operator presses the sampling switch (pin 1) when the component is in position, it will also be diverted using the piston
- When the component is diverted, a checking light (pin 2) flashes on and off 6 times over 3 seconds
- This system continually repeats.

Draw the flowchart to show the quality-assurance checks.

DON'T FORGET

If you run out of space in the question paper, create an arrow going back to the top and start again from there, as shown in the example. Make sure you have arrowheads, though, to show the direction of the flow.

ONLINE TEST

Test your knowledge of this topic by taking the test on the Digital Zone.

ELECTRONICS AND CONTROL

PROGRAMMABLE CONTROL: WRITING CODE WITH PBASIC 1

DON'T FORGET

Writing code **will not** feature in the question paper, but you do need to ensure that you are capable of doing it, as it will appear in your coursework.

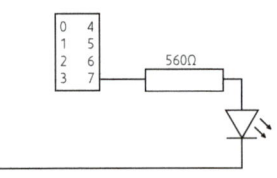

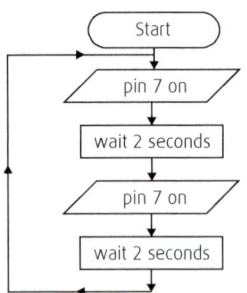

WRITING AND UNDERSTANDING CODE

Within your learning of programming microcontrollers, your school has the option to teach you any language they wish. There are several different types of microcontrollers available on the market today, and each has its own high-level language. What your school prefers will determine what you learn. Do not worry if you have been taught differently from any friends you have in other schools, as they all do the same things – nobody will be at a disadvantage because they are using a particular microcontroller.

Within the course, you will have to write code to ensure you are achieving the necessary outcomes, but the language you use to do this is completely up to your school. This section will cover one of the main languages currently used, namely PBASIC, through the use of Stamp controllers.

After you have produced the flowchart to any given specification, you must then work out the coding. For example, in the circuit shown here, the LED is to flash on for 2 seconds, then go off for 2 seconds. This will then continually repeat.

Input connection	Pin	Output connection
	7	LED

Firstly, we have to tell the microcontroller which pins are outputs, and which pins are inputs. We do this by using the 'let dirs' command. The board will have 8 pins that range from 0 to 7, and is read from right to left.

In this case, we are only concerned with one pin, and this is pin 7, and we want this to be an output. When stating this, we still have to set the nature of the other pins. In most cases, pins 0–3 will be inputs, and pins 4–7 will be outputs. To do this, we state in our first line of code:

let dirs = %11110000

1 is seen as being an output, and 0 is seen as being an input, and we read this from right to left.

let dirs = %1111 0000
               pin 7 pin 4 pin 3 pin 0

DON'T FORGET

Depending on the version of software that you use, it may not recognise this part of the code. If it doesn't work, just delete this line.

Next, it is good practice to set up symbols to name the pins you are using. This makes it easier to identify each pin when creating more complex programs, so it's worth getting into this habit now. In this case, we are setting up an LED connected to pin 7, so we will rename pin 7 'LED'.

After this, a label can be created for the program. This is a standard thing that is needed to allow it to loop back to this point. It doesn't matter what you call it – although it makes sense to call it something suitable or related to the program.

When switching the inputs or outputs on/off, we refer to the state of the signal. If it is on, it is known as **high** (logic 1, or on). If we want it to be off, it is referred to as **low** (logic 0, or off).

When creating delays in the program, we use the 'pause' command. When utilising this command, the time is measured in milliseconds (1,000ths of a second).

e.g.   1 second = 1,000

        5 seconds = 5,000

contd

Electronics and Control: Programmable Control: Writing Code with PBASIC 1

Once we have completed this, we have to finish the program. If it was to end, we would just write the code 'end', and it would finish. As we know, though, most programs will not want this to happen. Instead, it is more likely that it is intended to loop back to the beginning of the code so that the process can be repeated. To do this, we use the 'goto' command, and we ask it to 'go to' our label.

This means our program should read:

let dirs = %11110000

symbol LED = 7

MAIN:

    high LED

    pause 2000

    low LED

    pause 1000

    goto MAIN

It is also good practice to have the English translation written next to this code to help you understand what each line of code is meant to be doing. This can be very useful when the program is extremely complex. This can be achieved by using an apostrophe. After this apostrophe, the English translation can be written next to it. The apostrophe lets the software know to now stop reading this line of code, as it is no longer in a coding language. Instead, it will then move onto the next line.

let dirs = %11110000	' Setting up the directory
symbol LED = 7	' define pin 7 with the name 'LED'
MAIN:	' label called MAIN
high LED	' LED is switched on
pause 2000	' delay of 2 seconds
low LED	' LED is switched on
pause 1000	' delay of 1 second
goto MAIN	' loops back to the label called MAIN

 ONLINE

To practise your coding at home, download the PICAXE editing software from the link on the Digital Zone. This will allow you to write code, check if it works, and simulate it.

 DON'T FORGET

Syntax is **extremely** important when writing code. 'Main', 'MAIN' and 'main' would all be seen as different things. So, if your code does not work, go back and check your syntax and spelling. This is very likely to be the cause of your code not working.

 ## THINGS TO DO AND THINK ABOUT

A set of temporary traffic lights is required for a system of roadworks that must follow this sequence:

- The red light goes on for 12 seconds.
- The amber light then comes on for 4 seconds.
- The red and amber lights go off, and the green light switches on for 14 seconds.
- The green light switches off, and the amber light goes on for 3 seconds.
- The amber light goes off, then the system repeats.

 ONLINE TEST

Test your knowledge of this topic at www.brightredbooks.net

Input connection	Pin	Output connection
	2	Red light
	1	Amber light
	0	Green light

Draw the flowchart for this sequence, and write out the code.

43

ELECTRONICS AND CONTROL

PROGRAMMABLE CONTROL: WRITING CODE WITH PBASIC 2

DON'T FORGET

When connecting a switch to a microcontroller, it is good practice to connect it in a voltage-divider circuit with a pull-down resistor. This resistor will ensure that the input to the component is always at 0 unless desired differently.

IF COMMAND

The 'if ... else' command is used when a decision box is needed in your flowchart. For example, **if** a switch is pressed, a light will go on; **else** the light will stay off.

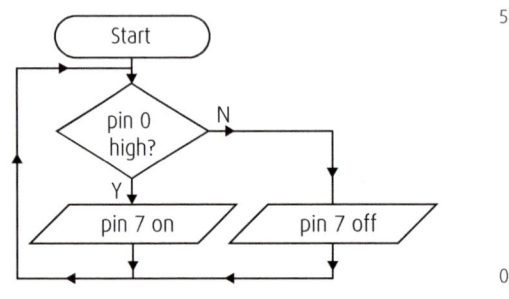
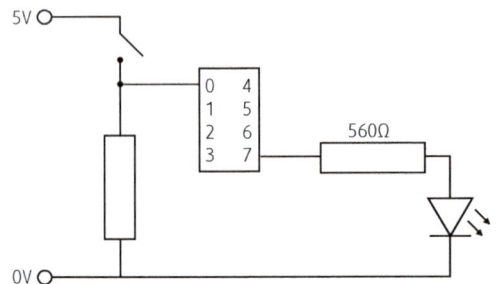

Depending on the hardware your school has, you may have to connect an input module to your Stamp controller to provide the interfacing circuits required to connect switches and sensors. When the slide switch on the input module is **up**, the input module provides four digital (on/off) switch connections. These can be used to connect input switches to the Stamp controller. The switches can then be connected through the screw-terminal blocks on the module.

Input pins 0 and 1 have on-board test switches. These allow programs to be tested without the need to connect external switches.

DON'T FORGET

Sub-procedures that complete common tasks can be copied from program to program. This will save time from having to constantly rewrite them.

To write the code in PBASIC, it is helpful to use labels. We already do this to begin the program, but further ones can be created to separate the program into smaller sections to make it easier to understand.

symbol led = 7	' define pin 7 with the name 'led'
MAIN:	' make a label called 'main'
if pin0 = 1 then LIGHT	' jump if the input is on
low led	' switch output 7 off
goto MAIN	' else loop back around
LIGHT:	' make a label called 'light'
high led	' switch output 7 on
goto MAIN	' jump back to start

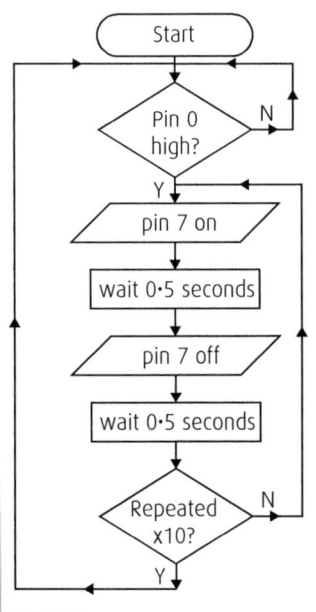

COUNTERS

It is often useful to repeat the same part of a program a number of times, for instance when flashing an LED. If the flowchart is amended so that the LED connected flashes on and off 10 times, we can see that the decision box is asking for a repetition.

To write the fixed-loop part of the code, it has to be stored in the memory of the microcontroller. This is done using variables. There are 10 variables, labelled b0 to b9, and any can be used for this purpose. It is also good practice to rename these variables using the 'symbol' command. This makes it easier to remember and understand what each command is meant to do.

contd

44

Electronics and Control: Programmable Control: Writing Code with PBASIC 2

symbol counter = b0	' define the variable 'counter'
symbol led = 7	' define pin 7 with the name 'led'
MAIN:	' make a label called 'main'
if pin0 = 1 then LIGHT	' jump to the LIGHT label if the input is on
goto MAIN	' else loop back around
LIGHT:	' make a label called 'light'
for counter = 1 to 10	' start a for ... next loop to loop 10 times
high led	' switch led on
pause 500	' wait for half a second
low led	' switch led off
pause 500	' wait for half a second
next counter	' end of for ... next loop
goto MAIN	' else loop back around

VIDEO LINK

Watch the videos on the Digital Zone to get a better grasp of how to program with PBasic.

THINGS TO DO AND THINK ABOUT

A car-manufacturing company is designing a heating system for the seats, and have decided that a microcontroller-based system should be used.

The heating system gives the option of three different settings: Off, Lo, or Hi.

Input Connection	Pin	Output Connection
	4	heater
	3	LED
Hi setting	2	
Lo setting	1	
Off setting	0	

Write the program for the flowchart shown in PBASIC.

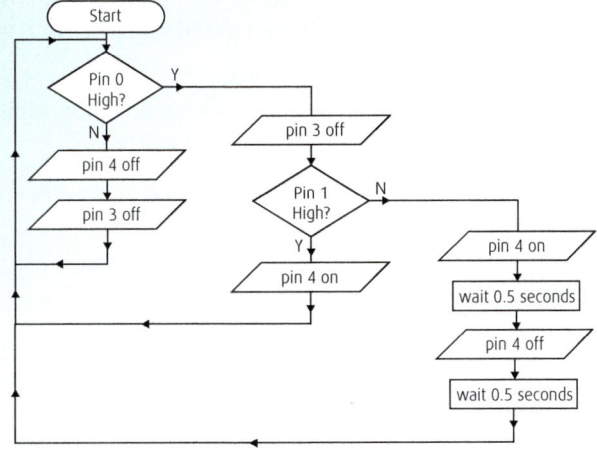

ONLINE TEST

Test your knowledge of this topic at www.brightredbooks.net

ELECTRONICS AND CONTROL

PROGRAMMABLE CONTROL: WRITING CODE WITH ARDUINO 1

WRITING AND UNDERSTANDING CODE

Within your learning of programming microcontrollers, your school has the option to teach you any language they wish. There are several different types of microcontrollers available on the market today, and each has its own high-level language. What your school prefers will depend on what you learn. Do not worry if you have been taught differently from any friends you have in other schools, as they all do the same things – nobody will be at a disadvantage because they are using a particular microcontroller.

Within the course, you will have to write code to ensure you are achieving the necessary outcomes, but the language you use to do this is completely up to your school. This section will cover one of the main languages currently used, and that is a variation on C++ through the use of an Arduino board.

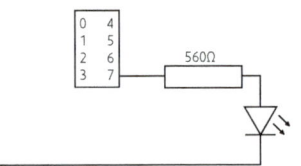

After you have produced the flowchart to any given specification, you must then work out the coding. For example, in the circuit shown here, the LED is to flash on for 2 seconds, then go off for 2 seconds. This will then continually repeat.

Input connection	Pin	Output connection
	7	LED

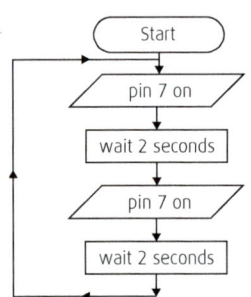

When using Arduino, you firstly have to set up integers to name the pins. This will make it easier to identify what each pin is when you reach more complex programs, so it's worth getting into the habit now. In this case, we are setting up an LED connected to pin 7, so we can use the command

int LED = 7;

We now have to set up two parts to the program that must **always** be there. These are the 'void setup()' and 'void loop()' commands.

void setup() commands are ones that will run when the Arduino first switches on. This is where you tell your board which pins are to be treated as inputs, and which ones as outputs.

void loop() commands are ones that will run continuously in a loop while the Arduino hardware is on. This is where your main program will be written.

Within the 'void setup()' is where pin 7 is set as an output, and the 'pinMode' command is used to set it as output.

```
void setup()
{
  pinMode(LED, OUTPUT);
}
```

Within 'void setup()' and 'void loop()' segments, the entirety of the section must be kept within parentheses (the curly brackets). It will not work if you do not do this. As it becomes more complex, there will be more brackets and parentheses being used, so you should ensure you have put them in properly, and in the correct place. To ensure this, click on one half of the parenthesis, and a box will appear around the other to show where the beginning and the end are.

Now that this is set up, the actual code needs to be written, and this will be written in the 'void loop()' area.

To switch a pin on, the 'DigitalWrite' command needs to be used. This needs two pieces of information – which pin you are to use (or what you have named it previously), and whether the pin will be HIGH (switched on) or LOW (switched off).

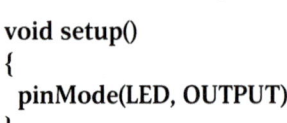

DON'T FORGET

After every line, you always have to add a semicolon, as this tells the software to move on to the next line of code.

```
void loop()
{
  digitalWrite(LED, HIGH);
}
```

A delay now needs to be put in to tell it how long this pin is to be on or off for. This is done using the 'delay' command. This delay is measured in milliseconds.

e.g.   1 second = 1000
        5 seconds = 5000 … etc.

This means our program should read:

```
void loop()
{
  digitalWrite(LED, HIGH);
  delay(2000);
  digitalWrite(LED, LOW);
  delay(2000);
}
```

It is also good practice to have the English translation written next to this code to help you understand what each line of code is meant to be doing. This can be very useful when the program is extremely complex. This can be achieved by using two forward slashes. After you have typed //, the English translation can be written next to it. The // lets the software know to now stop reading that line of code, as it is no longer in a coding language. Instead, it will then move on to the next line.

```
int LED = 7;                    // naming pin 7 'LED'
void setup()
{
  pinMode(LED, OUTPUT);         // making the LED an output
}
void loop()
{
  digitalWrite(LED, HIGH);      // switching on the LED
  delay(2000);                  // pausing the program for 2 seconds
  digitalWrite(LED, LOW);       // switching off the LED
  delay(2000);                  // pausing the program for 2 seconds
}
```

Arduino **always** creates a loop in the program (hence the name 'Void loop'), so it would then link back to the top of this section and switch the LED on again and continue to follow the sequence.

 ## THINGS TO DO AND THINK ABOUT

A set of temporary traffic lights is required for a system of roadworks that must follow this sequence:

- The red light goes on for 12 seconds.
- The amber light then comes on for 4 seconds.
- The red and amber lights go off, and the green light switches on for 14 seconds.
- The green light switches off, and the amber light goes on for 3 seconds.
- The amber light goes off, then the system repeats.

Input connection	Pin	Output connection
	2	Red light
	1	Amber light
	0	Green light

Draw the flowchart for this sequence, and write out the code.

 ONLINE

To practise your coding at home, sign up to https://www.tinkercad.com/, as this allows you to fully simulate Arduino. This will also allow you to practise your breadboard electronics.

 **DON'T FORGET**

Syntax is **extremely** important when writing code. 'Main', 'MAIN' and 'main' would all be seen as different things. So, if your code does not work, go back and check your syntax and spelling. This is very likely to be the cause of your code not working.

 ONLINE TEST

Test your knowledge of this topic at www.brightredbooks.net

ELECTRONICS AND CONTROL

PROGRAMMABLE CONTROL: WRITING CODE WITH ARDUINO 2

DON'T FORGET

When connecting a switch to a microcontroller, it is good practice to connect it in a voltage-divider circuit with a pull-down resistor. This resistor will ensure that the input to the component is always at 0 unless desired differently.

IF COMMAND

The '**if … else**' command is used when a decision box is needed in your flowchart. For example, **if** a switch is pressed, a light will go on; **else** the light will stay off.

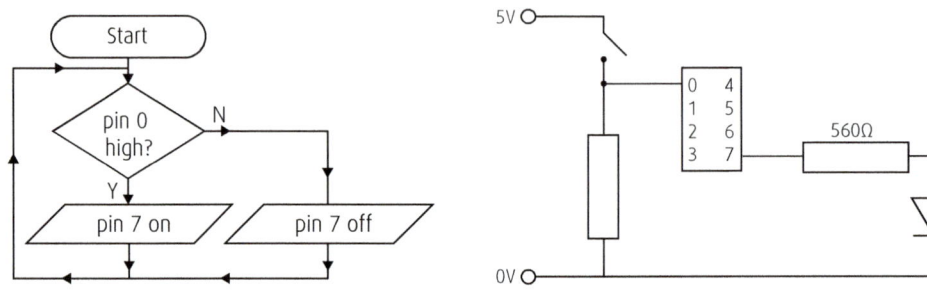

To write code in Arduino, the 'digitalRead' command is used. This allows the program to start to read the inputs and see what they are doing, and allows it to complete the program **if** the desired input has been achieved – **else** it does the other option stated.

```
int SWITCH = 0;
int LED = 7;
void setup()
{
  pinMode(SWITCH, INPUT);
  pinMode(LED, OUTPUT);
}
void loop()
{
  if (digitalRead(SWITCH) == HIGH)
        {
            digitalWrite(LED, HIGH);
        }
  else
        {
            digitalWrite(LED, LOW);
        }
}
```

DON'T FORGET

When using the 'if' command, it must use double equals sign. = means you are stating that one thing equals the other. == checks whether it is true or false.

COUNTERS

It is often useful to repeat the same part of a program a number of times, for instance when flashing an LED. If the flowchart is amended so that the LED connected flashes on and off 10 times, we can see that the decision box is asking for a repetition.

To create a finite loop in Arduino, we use this block of code:

{ for (int counter = 1; counter <= 10; counter = counter +1)

There are three separate statements in this command that are all separated by a semicolon. The first statement is the initialisation of the counter variable.

int counter = 1: this tells the Arduino board to start at 1 when it starts counting. This may seem an obvious starting point – but remember that a computer does not have common sense like we do. It only knows **exactly** what we tell it.

contd

Electronics and Control: Programmable Control: Writing Code with Arduino 2

counter <= 10: this section tells the computer the number of times it has to repeat this process. In this case, it has to count to 10.

counter = counter +1: this line of programming tells the computer how much to add to the value of the pin each time it repeats. In other words, every time the program completes a loop, it will add 1 until it reaches the set maximum.

After this command is written, the 10x flashing LED aspect of the program can be written. This has to be contained in { } parentheses. The reason for this is that the Arduino reads the program one line at a time. Without the brackets to tell it that this is a whole block of code, it will only repeat the next line. If the brackets are there, the microcontroller recognises that there is a whole section of code to read and repeat.

There should be no semicolon at the end of this command. This is because the counter section is seen as one whole command – if the semicolon was there, it wouldn't recognise what it is supposed to repeat.

```
int SWITCH = 0;
int LED = 7;
void setup()
{
  pinMode(SWITCH, INPUT);
  pinMode(LED, OUTPUT);
}
void loop()
{
  if (digitalRead(SWITCH) == HIGH)
        { for (int counter = 1 ; counter <= 10; counter = counter +1)
                { digitalWrite(LED, HIGH);
                  delay (250);
                  digitalWrite(LED, LOW);
                     delay (250);
                }
        }
else
        {
        digitalWrite(LED, LOW);
        }
}
```

 ONLINE

Visit www.brightredbooks.net for a link to the Arduino homepage, where you can up-skill your knowledge of programming with tutorials and examples.

 **DON'T FORGET**

You will note that white space (extra blank lines) and tabbing (moving lines of text along in line with each other) have been utilised in the example code to clearly show all the commands that are contained in the program. Breaking up the program like this can make it easier to read and understand what each section of it is.

 ONLINE TEST

Test your knowledge of this topic at www.brightredbooks.net

 THINGS TO DO AND THINK ABOUT

In order to give more control, and the ability to adapt the system in the future, a microcontroller is chosen for controlling the opening and closing of automatic toilet doors on a train. The specification is as follows:

- Doors open if either the *Inside* or the *Outside* button is pressed.
- A motor controlling the doors turns forward for 5 seconds.
- A *Warning* LED flashes 10 times over 2 seconds.
- Motor turns to close the doors.
- If *Sensor* is pressed during the closing action, then return to step 2.
- When sensor *Closed* is triggered, the system should return to the start of the sequence.

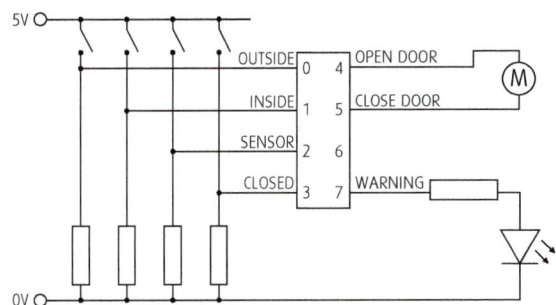

   a) Develop a flowchart to meet the given specification.

   b) Develop a program to meet the given specification.

   c) Describe how you would test the system.

   d) Suggest ways this design could be improved in the real world.

ELECTRONICS AND CONTROL

PROGRAMMABLE CONTROL: MOTOR CONTROL

HIGH-POWERED CONTROL

A microcontroller is only powered by a small voltage supply, and this is not enough to control of a high-powered device like a motor. To resolve this, a MOSFET can be used, as this is a driver suitable for larger loads with high currents.

As you know, a MOSFET essentially works in the same manner as a transistor, except it is operated by voltage instead of current. To use it within a circuit, it must be connected so that our high-powered device is connected to the supply voltage (V_{CC}) but not to ground (0 V). The ground instead is connected to the MOSFET's drain. This means that when our microcontroller sends a HIGH signal to the gate, it saturates the MOSFET (connecting the drain and source) and completes the circuit, allowing the high-powered output device to switch on.

As you may notice, a diode has also been put into the circuit parallel to the motor. Any time you are powering a device with a coil, such as a relay, solenoid or motor, you need this to protect the circuit. When the power is removed from the coil, a very high reverse voltage called EMF spikes back. This may only last a few microseconds, but it is long enough and powerful enough to destroy the driver. By adding this diode, it stops the EMF feeding back into the circuit and hence protects the MOSFET.

DON'T FORGET

Make sure that the protection diode is facing the correct way. The point of the triangle – or, if you are building it, the stripe – should be facing the V_{CC} of the device. If it is facing in the wrong direction, the device you are trying to power will not work, as the diode will just allow the current to bypass it.

PULSE-WIDTH MODULATION

When using motors, it is extremely important that you can control how fast the motor turns. There are two ways in which this can be done. The first is by far the simpler way – and that is to just vary the voltage applied to the motor. For example, if 2V is applied to a small DC motor, it will turn at a slower speed than it would if 5V was applied. The major drawback of this is that the torque of the motor will also drop. This means the motor will therefore be less powerful.

The second, and the more commonly used way, is to control the speed of the motor using **Pulse-Width Modulation** (PWM). With PWM, the full voltage is always applied across the motor, but it is rapidly switched on and off. As the voltage supply is off for some of the time, the motor is unable to reach its full speed. This is done so quickly that it cannot be seen by the human eye. So, instead of seeing the motor speeding up and slowing down, it will just seem that the motor is turning constantly at a different speed. The advantage of this speed-control system is that the torque does not drop, meaning the motor's power stays the same.

contd

VIDEO LINK

To see PWM and the analogWrite command in action, visit www.brightredbooks.net

Electronics and Control: Programmable Control: Motor Control

The time that the power supply is switched on is called the **mark** time, and the time that the motor is switched off is called the **space** time. By varying the mark-to-space ratio, the speed of the motor can be changed.

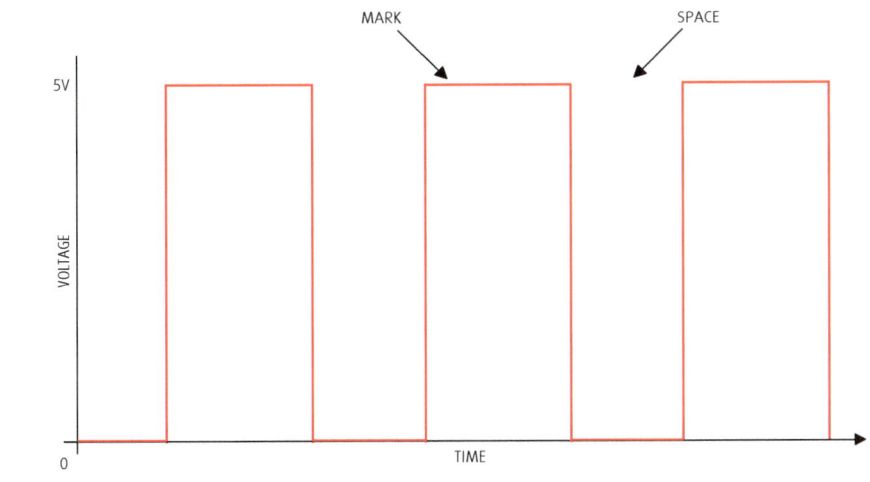

ONLINE

If you enjoy woking with Arduino, visit the Digital Zone to get some projects to build.

CREATING A FLOWCHART

When creating a flowchart for a soft start system, the flowchart doesn't have to be written out for each different section of the speed control. For example, a sub procedure is to be designed with the following specification:

- When a switch is pressed, a motor is Pulse Width Modulated in a 10-times repeat loop to build up its speed.
- Initially the MARK-SPACE Ratio is 2:10
- Each time around the loop the MARK is increased by 1 and the SPACE decreased by 1
- The system will then stop

Instead of drawing out a flowchart that says 'this pin is on for a certain time, off for a certain time, one again for a time, then off … etc.', the Mark and Space values can be stated at the beginning, and then altered throughout a fixed loop.

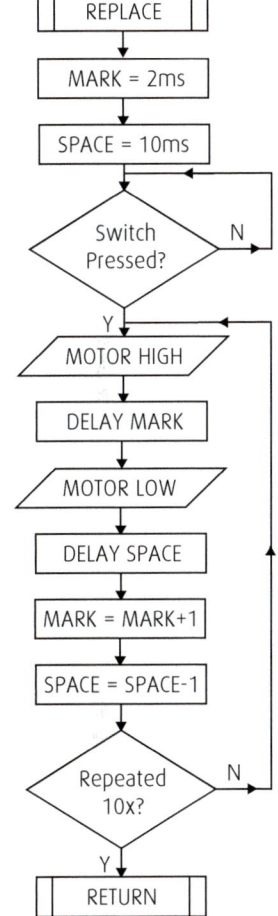

THINGS TO DO AND THINK ABOUT

At a computer game factory, a circuit board within the system undergoes a range of tests by applying signals to it. A microcontroller is used to provide these signals. The sequence of one of the test procedures is shown below.

- When the test 1 button is pressed the signal goes high for 500ms and then goes low.
- When the test 2 button is pressed the signal is pulse width modulated in a 10-times repeat loop.
- Initially the mark-space ratio is 12:2
- Each time around the loop the mark is decreased by 1 and the space increased by 1
- The system will then reset ready to repeat this sequence.

Draw a flowchart for this sequence.

ONLINE TEST

Test your knowledge of this topic at www.brightredbooks.net

51

MECHANISMS AND STRUCTURES

MECHANICAL SYSTEMS: DRIVE SYSTEMS

GEARING SYSTEMS

Gears are toothed wheels which are designed to transmit rotary motion and power from one part of a mechanism to another. They can be used within different gearing systems not only to change the direction in motion of the output, but also to increase or decrease the output speed of a mechanism. In this course, you will find it beneficial if you have an understanding about the following gearing systems.

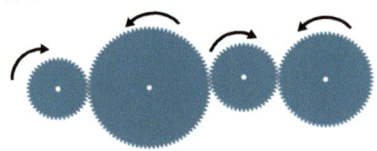

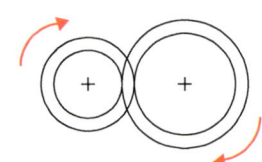

Simple Gear Train

Gears work by interlocking or meshing the teeth of the gears together as shown. When two or more gears are meshed, they form a system known as a **simple gear train**. The input gear which causes the system to move is called the **driver**. The output gear is known as the **driven**.

Idler Gear

An **idler gear** is a gear that can be inserted into a system, with its main purpose being to allow the driver gear and the driven gear to rotate in the same direction. It has no effect on the speed of the system.

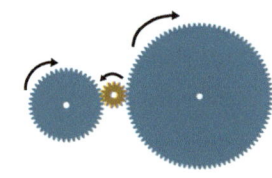

Compound Gear Trains

A **compound gear system** is created when multiple gear trains are connected by a common shaft. These are used to create a very large change in speed and are useful when there is only a small space available for the gearing system.

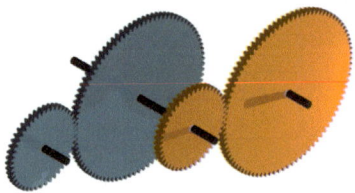

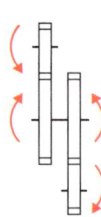

Worm and Wheel

Using a **worm and wheel** is a way of making a large speed reduction in a system as well as creating a large increase in torque. The worm (the part that looks like a screw thread) is usually fixed to a driver shaft, or even directly to a motor shaft. It then meshes with a wheel that is fixed to the driven shaft. This also allows the system to be rotated at 90° to the driver.

The worm wheel is seen as only having one tooth. This allows a huge reduction in speed, while taking up very little space. It also stops the gears from slipping and increases the torque.

Most guitars and other string-based instruments use a worm gear for the tuning mechanism. The gear's force reduction is the main reason for this, coupled with the locking capability that keeps the desired string tightness in place.

Worm and Nut

The **worm-and-nut** gearing system is used to convert **rotational** movement into **linear** movement. The worm gear is fixed so that when it spins, it moves the block left/right or up/down. This transmits the motion through the gear and is used in systems such as steering blocks in modern cars. The steering wheel rotates, causing the worm to turn. As this happens, the nut moves either to the right or left, depending on which way you turn the wheel.

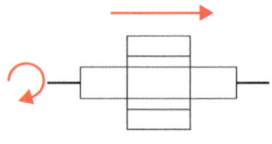

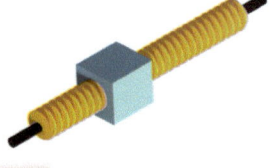

> **DON'T FORGET**
>
> The calculations to work out the gear ratios and speeds may also be needed within this course. To revise this area of the course and better your understanding of how the drive systems work, read back through the 'Mechanisms and Drive Systems' section in the National 5 Engineering Science Study Guide.

contd

Mechanisms and Structures: Mechanical Systems: Drive Systems

The worm-and-nut gearing system also allows for a big change in speed and increased torque.

Rack and Pinion

A **rack and pinion** is another gearing system that transfers rotational movement into linear movement. As the pinion (circular gear) rotates, it causes the rack (the flat panel with teeth) to move in a straight line. Rack-and-pinion gearing systems are often used as part of a simple linear actuator, and are used frequently in things like stair lifts. When the button is pressed and the motor turns, it rotates a gear which is connected to the rack. As the gear turns, it slides the seat along the rack, allowing the person to travel up or down the stairs.

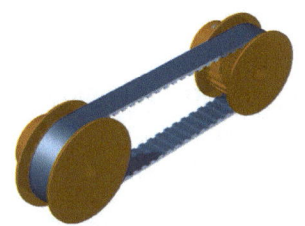
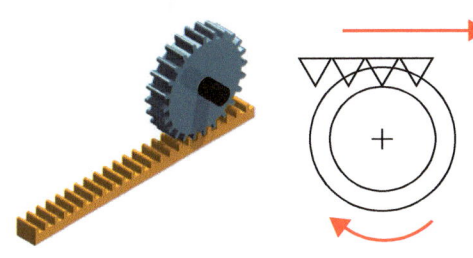

Belt Drive

To render a drive system useful, the motion often has to be transmitted from one part of a machine to another. Doing this through gearing systems is not always suitable, though, as connecting too many gears together can result in huge efficiency losses due to the friction created. A simple way of getting around this is by using a belt-drive system. The motion can then be transferred over a large distance by wrapping a belt around two pulleys.

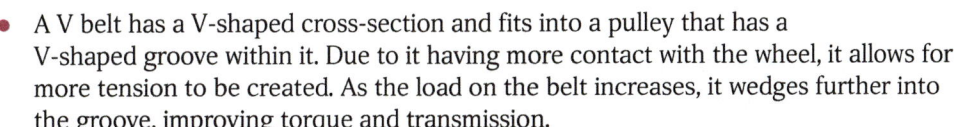
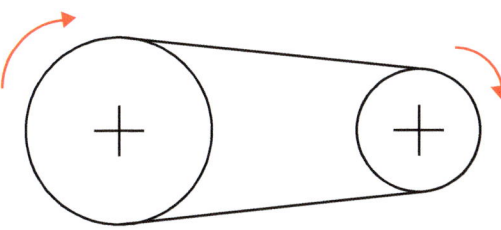

V belt   flat belt   toothed belt

There are three different types of belt that can be utilised – a V belt, a flat belt and a toothed belt.

- A V belt has a V-shaped cross-section and fits into a pulley that has a V-shaped groove within it. Due to it having more contact with the wheel, it allows for more tension to be created. As the load on the belt increases, it wedges further into the groove, improving torque and transmission.

- A flat belt has a rectangular cross-sectional shape. Its main advantage is that it allows slip when needed: for example, a motor system may cease because a gear won't turn. If this was a gearing system, it would continue to try to turn, eventually damaging the motor. Within a belt-drive system it would slip, stopping any damage. Its major disadvantage is that it also has a tendency to slip when a large load is applied, as there is not enough friction created between the belt and the pulleys.

- Toothed belts are like flat belts, but have 'teeth' like a gear. These belts do not allow slippage and are used when slippage in a system could cause damage.

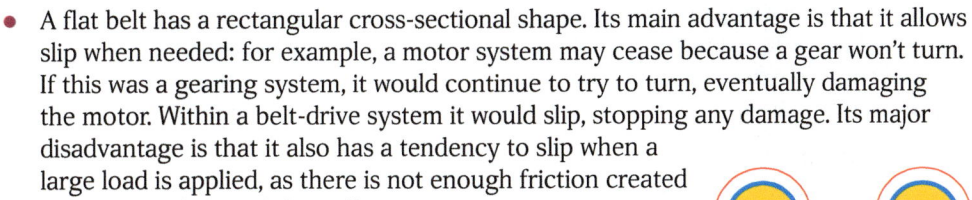

Another advantage of a belt-drive system is that changes in direction are also possible by crossing over the belts.

Chain Drive

The disadvantage of most belt-drive systems is that, even when properly tensioned, they can slip. This can be resolved by using a chain drive.

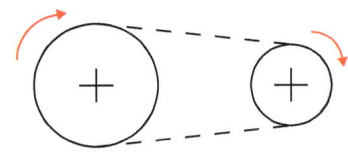

ONLINE TEST

Test your knowledge of this topic at www.brightredbooks.net

ONLINE

Visit the BrightRED Digital Zone to obtain a greater understanding of different gearing systems.

ONLINE

To learn more about gearing systems, head to www.brightredbooks.net

THINGS TO DO AND THINK ABOUT

Create a crossword for all the mechanisms you have learned. Give it to one of your classmates – can they solve it using the clues you have given them?

MECHANISMS AND STRUCTURES
MECHANICAL SYSTEMS: COUPLING METHODS AND CLUTCHES

COUPLINGS

Mechanical devices employ a variety of methods of transmitting motion or torque from one area to another, but mostly it is done through metal rods known as **shafts**. These shafts must be connected together to transmit motion – and this is done by using a device called a **coupling**. Couplings are used in machinery for several purposes, and there are many different types to suit any particular job. The most common uses of a coupling are:

- Connecting two different shafts together at their ends with the purpose of transmitting power, torque or movement.
- To provide the connection of shafts that are manufactured separately, such as a motor and generator, and to then allow disconnection for repairs or alterations.
- To introduce mechanical flexibility if the shafts are misaligned.
- To reduce the transfer of shock loads from one shaft to another.
- To alter the vibration characteristics of rotating units.
- To connect the driving and the driven parts.

Rigid Couplings

When shafts are perfectly aligned with each other and the shafts cannot move, they are joined with one of two types of coupling known as **flange** couplings or **muff** couplings. These couplings must be **keyed** to the shafts they are joining to give a positive drive.

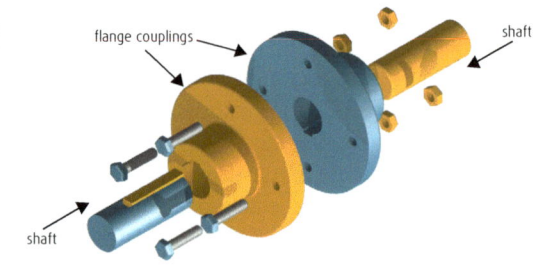

The **flange** coupling consists of discs which are fixed at the end of each shaft. The two flanges are then bolted together to complete the drive.

The **muff** coupling acts like a sleeve, covering both shafts, and once again is bolted together to hold the two shafts together.

Aligned couplings are most commonly employed in devices that have a medium- to heavy-duty load with moderate speed, like large turbine generators, and in certain process machines where the timing of the operation must be kept to exact standards, meaning that the shafts cannot be misaligned, as it would slightly alter the rate of turning.

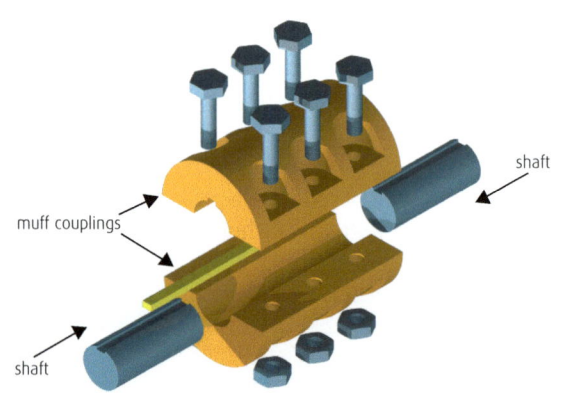

> **DON'T FORGET**
>
> It should be expected that you will be asked about coupling methods within any assessment. It may be worth writing a bullet-point summary of this double page spread and memorising it so you can be sure you fully understand how a clutch is used within a drive system.

> **VIDEO LINK**
>
> Watch the animation at www.brightredbooks.net to see exactly how a clutch works.

contd

Flexible Couplings

When the shafts in a system are misaligned, **flexi couplings** are frequently used, as they can accommodate the flaws and dynamics inherent in most systems, by countering the vibration that shafts may encounter, or compensating for misalignment if the shafts meet at a slight angle. This is very similar to a flange coupling, but has an additional rubber section inserted between the two couplings. Rubber's material properties mean that it can be 'squashed', allowing it to compensate for any small changes in the angle.

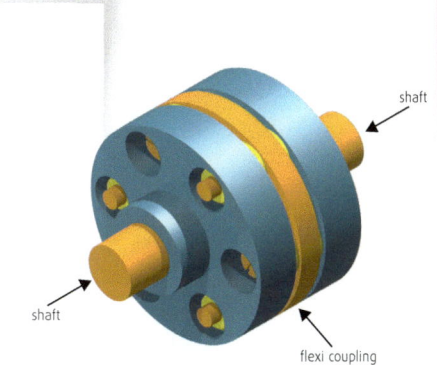

These couplings are most commonly employed in devices to protect shafts against the harmful effects from shock loads, hard starts of the motor, vibrations, or machines that may heat up and cause thermal growth of the material.

When the alignment of the shafts is more than a few degrees out, a universal joint must be used. A universal joint can transmit motion through an angle of up to 20°. The two shafts are attached to a **yoke**, and these yokes are free to pivot around a central part known as a **spider**.

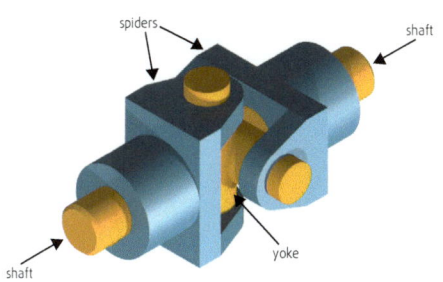

This form of coupling is most commonly used in applications such as drive shafts, automotive propeller shafts or control mechanisms, where space is an issue and the matter of misaligned shafts cannot be resolved in any other way.

CLUTCHES

A clutch is a special type of coupling that allows two rotating shafts to be connected and disconnected. It is used to disconnect the driver from the driven aspects of the mechanical system, and then to connect it back together again. For example, in a car, when the clutch is pressed, it disconnects the wheels from the engine. This will allow the car to roll free; and gear changes can be made without damaging the engine. The clutch is also pressed when you need to brake, as it allows the engine to keep turning but the wheels to stop.

Although there are huge advantages to using a clutch in a system, there are several different disadvantages. Firstly, as the system is now more complex, there is now a higher chance of things going wrong due to the increase in complexity and the increase in parts used. This means that regular maintenance is required due to wear, and the system potentially could be more expensive because of this. A clutch also has the potential to slip, which means energy could be wasted.

A clutch uses friction to its advantage in the way it operates. When connected, the two plates are pushed together so tightly that the driver rotating shaft will cause the driven shaft to rotate, causing a transmission of speed and power.

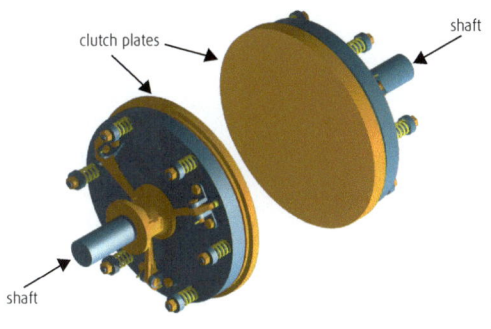

An example of this would be an electric drill. One shaft would be driven by the motor, and the other would be connected to the drill chuck. When switched on, the clutch plates connect, joining the two shafts together, ensuring they spin at the same speed and ensuring transference of power. A clutch can slip, though, which means that when the drill comes up against a large resistance, the plates will still be locked together, but spinning at different speeds. This ensures there is still torque within the system.

This issue can be resolved with the use of dog clutches or multi-plate clutches.

 ### THINGS TO DO AND THINK ABOUT

Create a crossword on couplings and clutches, and the different types that exist. Give it to one of your classmates – can they solve it from your clues?

 ONLINE TEST

Test your knowledge of this topic at www.brightredbooks.net

MECHANISMS AND STRUCTURES

MECHANICAL SYSTEMS: FRICTION

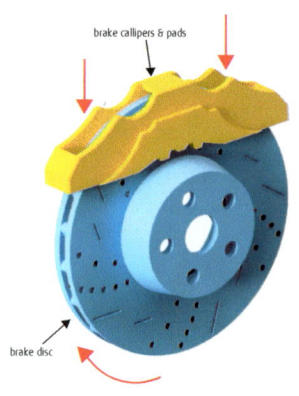

brake callipers & pads

brake disc

FRICTION

Friction is something that happens in all mechanical systems. It can be used to an engineer's advantage, for example in a braking system. If the force is applied to a moving part, for example a brake disc, it will cause it to slow down. If this applied force from the brake calliper and pad becomes greater than the turning force of the brake disc, it will cause the system to stop.

In complex mechanical systems, friction can also be an unwanted problem. As different parts move and touch each other, it creates heat. This potentially could be a fire risk, or cause the components to expand, distort or change shape. In turn, this will disrupt or destroy your system.

To reduce this unwanted friction, two methods can be employed: the lubrication of any moving parts, or installing bearings.

DON'T FORGET

If you have to refer to lubrication in a system, make sure you are being specific. Unless you are stating **exactly** what is to be lubricated, you will lose the marks!

LUBRICATION

Using some form of lubricant, such as an oil or grease, can help to reduce this unwanted friction. By lubricating the moving parts, it allows the different parts to touch and move more easily without creating so much heat.

BEARINGS

Bearings are components that are designed to withstand wear. Essentially, bearings are used to help support the weight of the system while allowing something to rotate, for example a wheel. They can also be used to replace any rubbing with rolling actions instead, as this rolling reduces the contact surface of the two touching components and therefore reduces the friction within a system. As the friction is reduced, it will therefore also reduce the energy loss in the system, as well as saving wear on the shaft. This means that the system will last longer, and, when it does eventually wear down, only the bearing will need to be replaced and not the whole mechanical system.

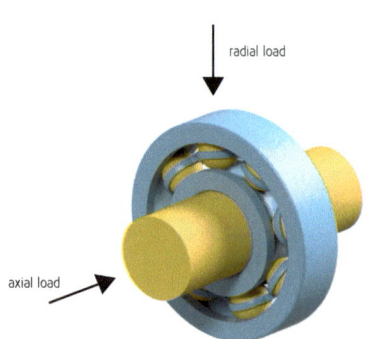

radial load

axial load

There are many different types of bearing, and they can be used for a variety of purposes, but one of the factors that would help decide the type of bearing is the type of external force that will be applied to it. A bearing can receive either an axial load or a radial load. A radial load is one that is perpendicular to the shaft, whereas an axial load is one that is parallel to the shaft. A good example of an axial load is the forward thrust on boats or prop-driven aeroplanes as a result of their propeller's rapid rotation, whereas a radial load would be the weight of a car pushing down on the wheel bearing.

Plain Bearings

A plain bearing, also known as a **bush** or **bushing**, is the simplest type of bearing. It consists of just a simple bearing surface, and has no rolling element. The bearing just slides over the shaft to give a surface for rotary applications. These are mainly used in systems that have a sliding or rotating shaft component.

An example of its use would be within an outdoor rotary clothes-dryer. This would be placed in between the hole in the ground and the dryer's main pole, allowing it to rotate freely.

contd

Roller and Ball Bearings

Roller and ball bearings consist of an outer and an inner **race** which have grooves machined into them. Hardened steel balls or **rollers** are then fitted between these, allowing them to rotate freely. As the shaft rotates, it also rotates the rollers/balls, and because these are rolling they have a much lower friction rate than if two flat surfaces were sliding against each other. In use, the balls and rollers must be well lubricated, though, to ensure that a build-up of heat does not happen.

A ball bearing is used when working with small loads and high speeds, and is frequently used in things like skateboards and drills.

Roller bearings are more suited to dealing with high loads and lower speeds. Compared to ball bearings, they are better at supporting heavy radial loads but can only handle limited axial loads. They are commonly used in agricultural equipment or rotating solar panels, where the weight they are carrying is extremely heavy and should move slowly.

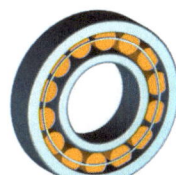

Thrust Bearings

Thrust bearings are a special type of bearing that is designed to support a high axial load. When a shaft has a large axial load, it must have a thrust bearing.

Thrust bearings are used frequently in things like office chairs, as they are expected to support not only high axial loads but also sudden high-impact loads when someone sits down on them.

Journal Bearings

A journal bearing is used when a radial load is applied to a mechanism. In thrust bearings, there is higher contact than there would be with a journal bearing, so wear occurs very quickly. In the case of journal bearings, the contact between the shaft and bearing is eliminated, as it is suspended within a bearing chamber. This means it will wear less and will increase the bearing's lifetime, and makes it more efficient.

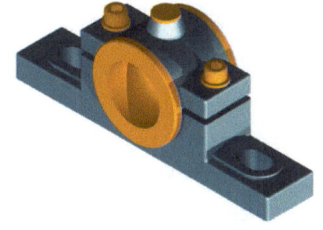

ONLINE TEST

Test your knowledge of this topic at www.brightredbooks.net

As journal bearings are designed to reduce load friction, they are often used when the load is light and the motion is continuous. Because of this, they are most often used in industrial machines that require high horsepower and high loads, like turbines or pumps.

Split Bearings

As bearings are designed to wear, it makes sense that they must eventually be removed and replaced. When the bearing support is at the start or end of a shaft, it is simple to access, and this replacement can be done easily. However, when a shaft is long, it may be supported at several points along its length, and be difficult to access. To make it easier to access the bearings, split bearings can be used, as they can be taken apart and reached more easily.

VIDEO LINK

For a greater understanding of bearings, watch the video at www.brightredbooks.net

Split bearings are used in a range of applications to reduce downtime, save on costs, and increase overall safety. They can be frequently found in places such as mixing and stirring areas in water-treatment facilities or the pulp and paper industry, where an extended period of downtime is undesirable.

THINGS TO DO AND THINK ABOUT

Take notes of this page and write a bullet-point summary as to why bearings would be used in a mechanical system, for example a motorbike.

MECHANISMS AND STRUCTURES

MECHANICAL SYSTEMS: TORQUE AND MECHANICAL POWER

TORQUE

Torque is the name of the force needed to make an object turn in a rotary motion. To calculate torque, the equation used is:

T = Fr

Torque = Force × radius

- **Force** is measured in **Newtons (N)**
- **Radius** is measured in **metres (m)**
- The unit for measuring **torque** is **Newton metres (Nm)**.

Example:

The shutter slats on a garage door are rolled up and down over a drum mechanism with a diameter of 0·5 m. The mass of the garage door is 70 kg.

Calculate the torque supplied to the drum mechanism to lift the garage door at a constant speed.

Show all working and final unit.

With a question like this, the first thing to do is use your data booklet (if you don't already know it) to find the equation for torque.

T = F × r

After this, put any information you may know into the calculation. Depending on the complexity of the question and how much it is worth, this could be worth 1 mark.

In this case, the force and the radius are both unknown quantities, so some other calculations will have to be done first to work these out values.

Force: the only information we have that is relevant is that the mass is 70 kg; but the force can be calculated by m × g.

F = mg
  = 70 × 9·8
  = **686 N**

Radius: The diameter of the drum is 0·5 m. The radius is half the diameter, so this must be 0·25 m.

Therefore:
T = F × r
  = 686 × 0·25
  = **171·5 Nm**
  = **170 Nm (to the same significant figures as the question)**

DON'T FORGET

Remember that the unit for distance is in metres, so you may have to change to this if the question asks in a different unit.

MECHANICAL POWER

The **mechanical power** is the power created as a system turns. This differs from torque as, although both are forces calculated from rotation, for power to be produced, movement over a distance must happen. A system with no rotation can deliver torque – like an electric motor – but, since no distance is moved by force, no power will be produced.

Mechanical power is calculated using one of the following equations:

P = Fv

Power = Force × velocity

contd

Mechanisms and Structures: Mechanical Systems: Torque and Mechanical Power

- **Force** is measured in **Newtons (N)**
- **Velocity** (or speed) is measured in **metres per second (ms⁻¹)**
- The unit for measuring **power** is **Watts (W)**.

OR

P = 2πnT

Power = 2 × 3·14 × the rotational speed per second × Torque.

- **n** is the **rotational speed per second (rev sec⁻¹)**
- **π** is 3·14
- **Torque** is measured in **Newton metres (Nm)**
- The unit for measuring **power** is **Watts (W)**.

VIDEO LINK

Watch the video at www.brightredbooks.net to obtain a greater understanding of torque versus power.

Example 1:

A racer is riding a bike at an average speed of 15 ms⁻¹. If they are riding into a headwind and burning energy at a rate of 500 J/s, assuming that 80% of this energy is going into overcoming the air resistance, calculate how much force the air is exerting on the racer.

The power used to overcome the air resistance is 80% of 500 W = 400 W.
P = Fv => F = P / v
= 400 / 15
= **27 N**

Example 2:

A fairground ride has six seating areas, which are 12 m from the centre of the ride, with each seating area subjected to 50 N of wind resistance.

Calculate:
i) The torque produced by the drive shaft to overcome the wind resistance.
ii) The mechanical power required by the motor if it is to complete 30 revolutions in 2 minutes.

i) Per car:
T = Fr
= 50 × 12
= 600 Nm

This now needs to be multiplied by 6, as there are 6 different seating areas:
T_total = 600 × 6
= **3,600 Nm**

ii) To work out the number of revolutions per second, we have to first calculate the number of revolutions it takes in seconds:
n = 30 / 2
= 15 rev min⁻¹
= 15 / 60
= **0·25 rev sec⁻¹**

P = 2πnT
= 2 × 3·14 × 0·25 × 3,600
= **5,652 W**
= **5·6 KW (to the same significant figures as the question)**

ONLINE TEST

Test your knowledge of this topic at www.brightredbooks.net

THINGS TO DO AND THINK ABOUT

a) A gearing system requires a force of 250 N from its final gear. Calculate the required torque if it has a diameter of 50 mm.

b) An electric motor is rated at 750 W and has a torque of 12·5 Nm. Calculate the speed of rotation of the motor.

c) A motor has a spindle that rotates at 1,000 rev min⁻¹ and has a power rating of 2·5 kW. Calculate the torque it produces.

MECHANISMS AND STRUCTURES

PNEUMATIC SYSTEMS: VALVES AND CYLINDERS

OVERVIEW

Pneumatics are a mechanical system that is very much like hydraulics, but instead of being powered by a fluid, compressed air is used as its source of energy. Pneumatic systems are commonly used in industry, as they are clean (there will be no mess if a leak occurs) and dependable (there is always air), and no sparks will occur that could cause a fire.

The equipment you will use in this area of engineering can be split up into two basic categories – **cylinders** and **valves**.

Cylinders are the 'muscles' of the pneumatic system, as they are used to move, hold or lift objects. They can even be used to operate other pneumatic components. Cylinders are operated by compressed air, and they convert the stored energy within it and transfer it into a linear motion.

Valves are used to control the flow of compressed air into a cylinder. They can be used either to turn the air on or off, or to change the direction in which the air is flowing.

3/2 VALVES

This component is known as a 3/2 valve – or, to be specific, a 'push-button actuated, spring-return 3/2 valve'. It gets its name because it has **3 ports** (3 connections) and **2 states** (on or off).

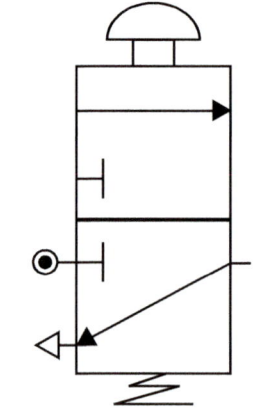

For this 3/2 valve to work, the push-button actuator must be pressed. When this is pressed, it essentially moves the top square down and replaces the bottom square. This will then create new connections to the mains air and to the output, allowing air to flow. As this is spring-return, when the button is released it will be pushed back up, reconnecting the output to the exhaust, and stopping the air from flowing.

3/2 valves can also be used to create a pneumatic version of an AND gate. By adding another push-button 3/2 valve, the first **and** second valves need to be pressed for the **single acting cylinder** to out-stroke.

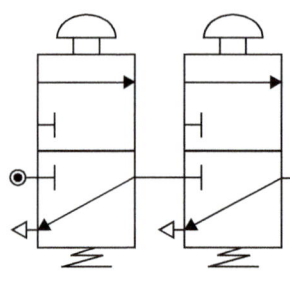

In addition to this, an OR gate can also be created with a 3/2 valve. By adding a component called a **shuttle valve**, this is now possible.

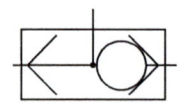

A shuttle valve is used to change the direction of air in the circuit. A small ball inside the component gets blown from side to side, allowing air to pass through to the cylinder from one side **or** the other. If Valve A **or** Valve B is pressed, the single acting cylinder will then out-stroke. Once any of the buttons is released, it will then in-stroke, as it's attached to a spring-return single acting cylinder.

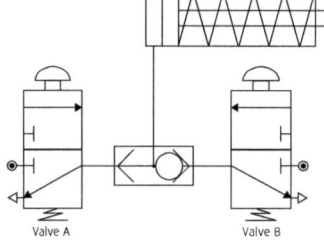

> **DON'T FORGET**
>
> When naming a component, remember always to name the types of actuators used in it.
>
>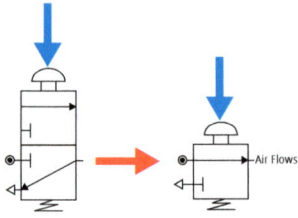

> **ONLINE**
>
> Head to the BrightRED Digital Zone to simulate some basic pneumatic circuits.

60

Mechanisms and Structures: Pneumatic Systems: Valves and Cylinders

5/2 VALVES

5/2 valves are needed when controlling a double acting cylinder. Although it is possible to do it with only two 3/2 valves, the pressure will be lost as soon as the valve is no longer pressed. This means the cylinder could then be easily pushed back/pulled out by hand and would be of no use in an industrial setting.

The 5/2 valve works in the same way as a 3/2 valve, except there are 2 exhaust ports and 5 output connections. A 5/2 valve is usually actuated by pilot air, as shown by a dashed line. Pilot air is a short burst of air that will activate the valve. In this case, you can see that a 3/2 valve is going to be used to switch on the 5/2 valve.

In this valve, if the right solenoid receives a signal, it will actuate the 3/2 valve and supply air from the right side. This will send pilot air into the right-hand side of the 5/2 valve and cause the double acting cylinder to in-stroke.

If the solenoid from the left-hand 3/2 valve receives a signal instead, it will actuate the left-hand 3/2 valve. In turn, it will send pilot air to the left-hand side of the 5/2 valve, sliding the left-hand section across and creating new connections to the output, mains air and exhaust. This will then cause the cylinder to out-stroke.

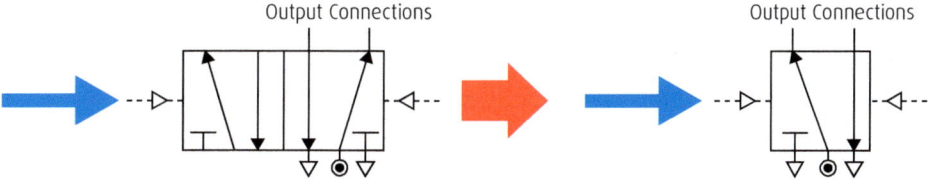

ONLINE

Visit www.brightredbooks.net to obtain a greater understanding of pneumatic components.

DON'T FORGET

Pilot air must always be used when a valve is to be used to switch on another valve.

THINGS TO DO AND THINK ABOUT

A pneumatic circuit is used to crush aluminium cans in a recycling plant. For safety reasons, the double acting cylinder should only out-stroke when a foot pedal and a push button have been pressed. It will then in-stroke when a roller trip has been actuated.

Draw the circuit that would complete this task.

ONLINE TEST

Test your knowledge of this topic at www.brightredbooks.net

MECHANISMS AND STRUCTURES

PNEUMATIC SYSTEMS: SPEED CONTROL AND TIME DELAYS

SPEED CONTROL

The problem with cylinders is that they can in-stroke and out-stroke very quickly. In a real-world situation, this could be very dangerous for the operator, and could damage the components due to wear and tear. To solve this problem, specific components can be used to control the flow of air through the valves. One way of doing this is using a **restrictor**.

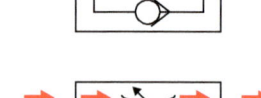

Using a restrictor will slow down the air in both directions, meaning that it will slow down the in-stroke as well as the out-stroke. This is essentially like putting a screw into the air pipe. If the screw is tightened, it creates a smaller area for the air to get through, which will therefore slow it down.

The main problem that exists with using this component is that sometimes the cylinder is to be slowed down in one direction only. This necessitates the use of a **unidirectional restrictor**.

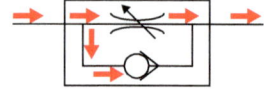

A unidirectional restrictor slows down the air in one direction only. The direction in which it slows air is dependent on the section with the cup and ball. As you can see, if the air flows through the component, it gets split between two paths. If the air flows from the direction facing the ball, the ball will be blown into the cup, blocking this path. This means there is now only one path for the air to go through, and this in turn will slow it down.

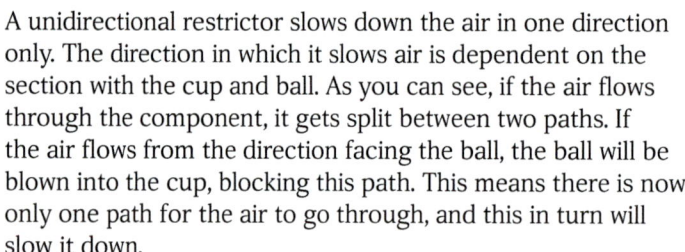

If the air flows from the other direction, the ball is instead blown away from the cup. This doubles the paths for the air to flow though, by allowing it through this second path. As the air flow is now unrestricted, the air will flow at full speed.

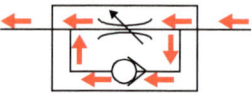

The unidirectional restrictor can be placed in the path of a cylinder to slow down either its in-stroke or its out-stroke. In this example, the out-stroke will be slowed down.

When valve A is pressed, the 5/2 valve will change state and start to supply the double acting cylinder with air, causing it to out-stroke. As the air is restricted on that right-hand path, the air that is already in the cylinder can only escape slowly. This will allow the cylinder to out-stroke not only slowly but also smoothly without affecting the force exerted.

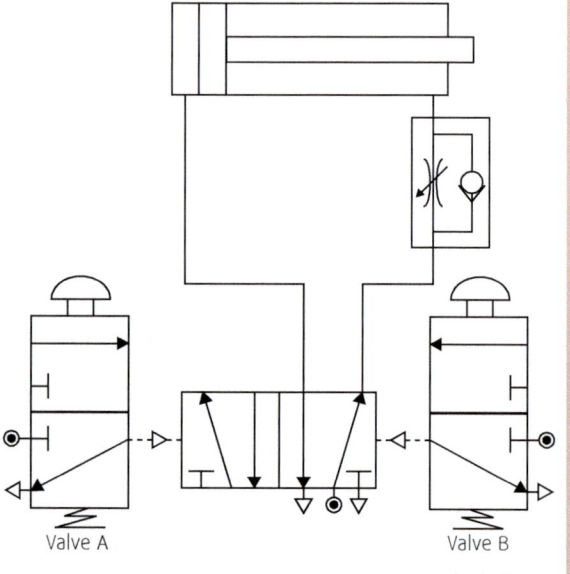

If the unidirectional restrictor is placed on the other pipe exiting the 5/2 valve, the cylinder will in-stroke slowly. In this example, when valve B is pressed, the 5/2 valve will change state and start to supply the double acting cylinder with air. This

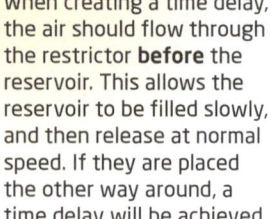

DON'T FORGET

When creating a time delay, the air should flow through the restrictor **before** the reservoir. This allows the reservoir to be filled slowly, and then release at normal speed. If they are placed the other way around, a time delay will be achieved, but the output air will be released slowly.

contd

will cause the cylinder to in-stroke slowly, as the air in the cylinder cannot escape as fast. Once again, doing this causes the cylinder to in-stroke not only slowly, but also smoothly and without affecting the force exerted.

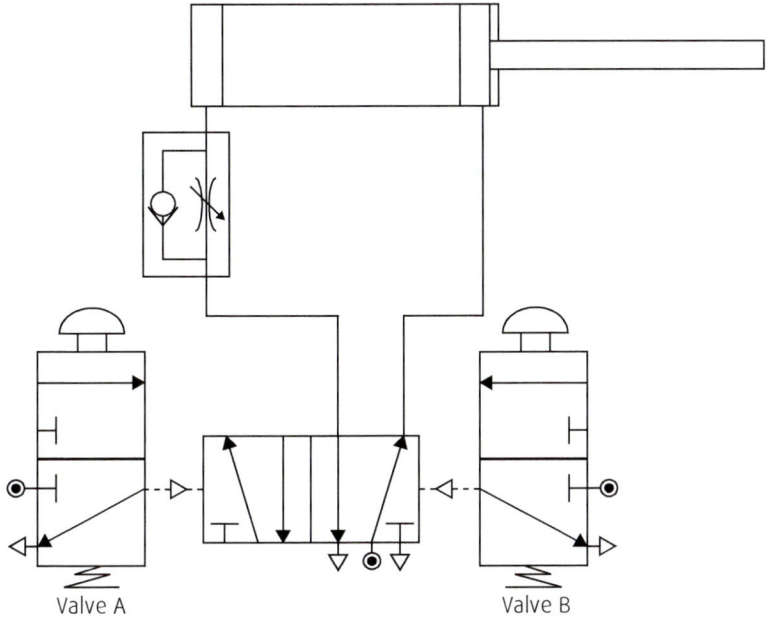

TIME DELAY

On many occasions, it is desired for there also to be a time delay between when the valve is actuated and when the cylinder is to respond. This pause can be created by a component known as a **reservoir**.

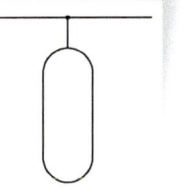

This is simply a container to store compressed air. By connecting this to a pipe, it increases the space that has to be pressurised before the next component can be operated. This will then create a time delay. To change the length of delay created by this component, it is as simple as changing the size of the reservoir.

A reservoir and a unidirectional restrictor are frequently used in conjunction with each other. The reason for this is that it will create a longer delay without the need for a huge reservoir, which could take up a lot of space. If the air is slowly filling up the reservoir, it makes sense that it will take longer for the signal to then be passed on to the next component.

VIDEO LINK

Watch the video on the Digital Zone to see how a time delay works within a system.

ONLINE TEST

Test your knowledge of this topic at www.brightredbooks.net

THINGS TO DO AND THINK ABOUT

Describe, using appropriate terminology, the operation of the circuit.

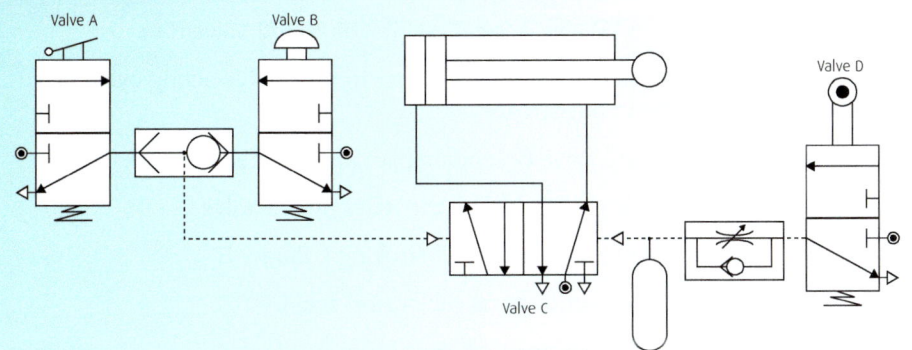

MECHANISMS AND STRUCTURES

PNEUMATIC SYSTEMS: SEQUENTIAL CONTROL

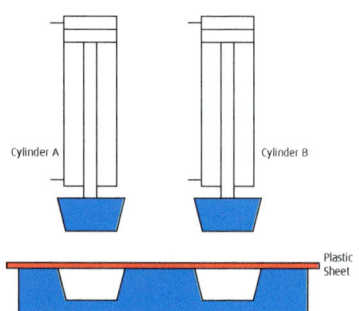

OVERVIEW OF SEQUENTIAL CONTROL

Pneumatic systems can be designed to perform multiple tasks, but more importantly they can be designed to complete tasks in a particular sequence.

For example, a company uses cylinders to press metal dies into hot plastic to create parts for a child's toy. As each cylinder is lowered individually, it presses the plastic into the shaped recesses.

The sequence of operations for this process is as follows.

- An operator pushes a button to start the process.
- Cylinder A lowers.
- Cylinder B lowers.
- Cylinder A rises.
- Cylinder B rises.

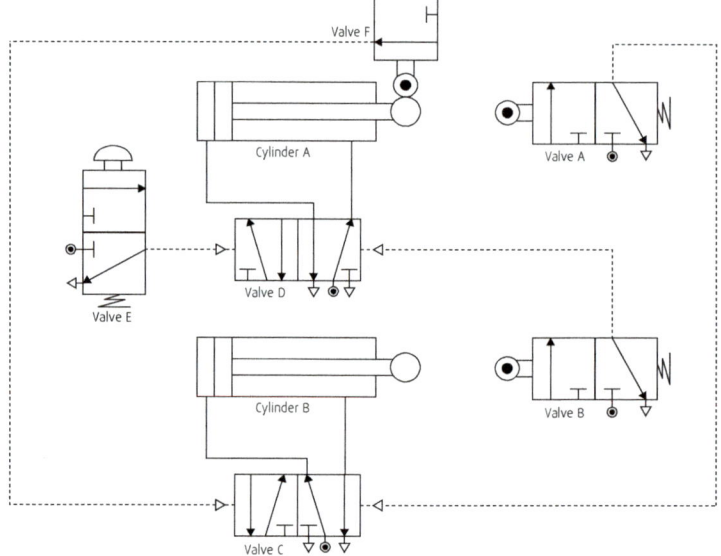

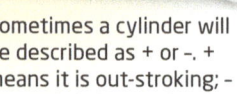

Sometimes a cylinder will be described as + or –. + means it is out-stroking; – means it is in-stroking.

In the exam, it is very probable that you will be asked to describe how a circuit like this works. If this happens, find the starting point by checking what section will need human interaction. On many occasions it will be at 'A' or '1', but like this example it is not always the case. When trying to answer the question, break it down into chunks, and follow the logical paths. An explanation for this circuit is given here to show you how this works.

If asked to describe a circuit like this in an exam, **always** start by finding the starting point, and follow the path. Use your finger to do this if needed! It is good practice to name all the components at least once each. If it is named, for example 'Valve A', put this in brackets after the name. If you return to this component, you can then just refer to it as 'Valve A'.

- When the push-button spring-return 3/2 valve is activated, it will change the state of 5/2 Valve A.
- This causes the double acting cylinder (Cylinder A) to out-stroke.
- This will hit the roller on 3/2 Valve A, sending pilot air to 5/2 Valve B.
- This causes the valve to change state, sending air to the double acting cylinder (Cylinder B), causing this to out-stroke.
- This will hit the roller on 3/2 Valve B, sending pilot air to 5/2 Valve A.
- This causes Cylinder A to in-stroke, hitting the roller on 3/2 Valve C.
- This in turn sends pilot air to 5/2 Valve B, in-stroking Cylinder B.
- The system now waits for a user to press the button again.

This could be extended even further and look even more complicated, but it is nothing you should ever panic about. If you have an understanding of how the components work, and you are able to see where to start, just follow the logical path and take it one part at a time.

Mechanisms and Structures: Pneumatic Systems: Sequential Control

For example:

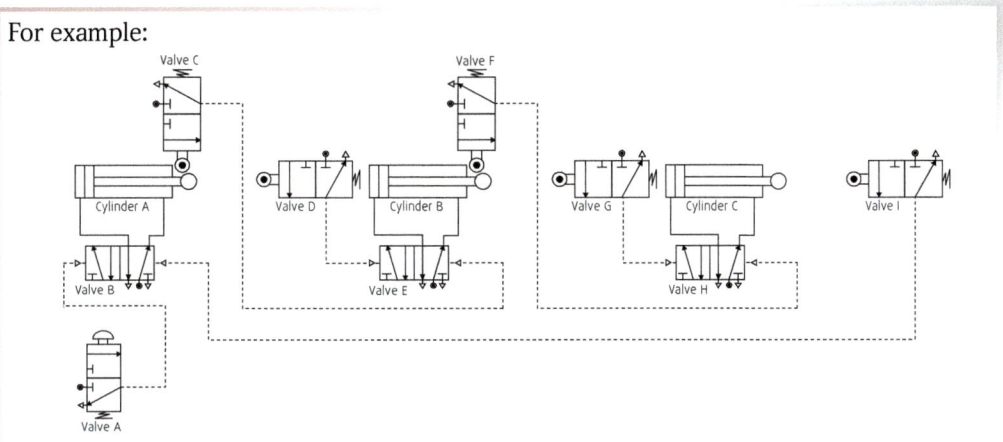

This looks a big and complicated circuit, but it can be seen that Valve A involves some human interaction, so it would suggest that this is an obvious starting point.

- The system will begin when someone actuates Valve A (the push-button spring-return 3/2 valve).
- This will then send pilot air to Valve B (the 5/2 valve), changing its state and causing Cylinder A to out-stroke.
- This will hit the roller on Valve D, sending pilot air to Valve A, causing it to change state.
- This causes Cylinder B to out-stroke and hit the roller trip on Valve G.
- This sends pilot air to Valve H, causing it to change state.
- This causes Cylinder C to out-stroke and hit the roller trip on Valve I.
- This sends pilot air back to Valve B, causing it to change state.
- This causes Cylinder A to in-stroke, hitting the roller on Valve C.
- This sends pilot air to Valve E, causing it to change state.
- This causes Cylinder B to in-stroke, hitting the roller on Valve F.
- This sends pilot air to Valve H, causing it to change state.
- This causes Cylinder C to in-stroke, hitting the roller on Valve F.
- The system is then reset, waiting for the user to manually push the button on Valve A.

THINGS TO DO AND THINK ABOUT

Describe, using appropriate terminology, the operation of the circuit.

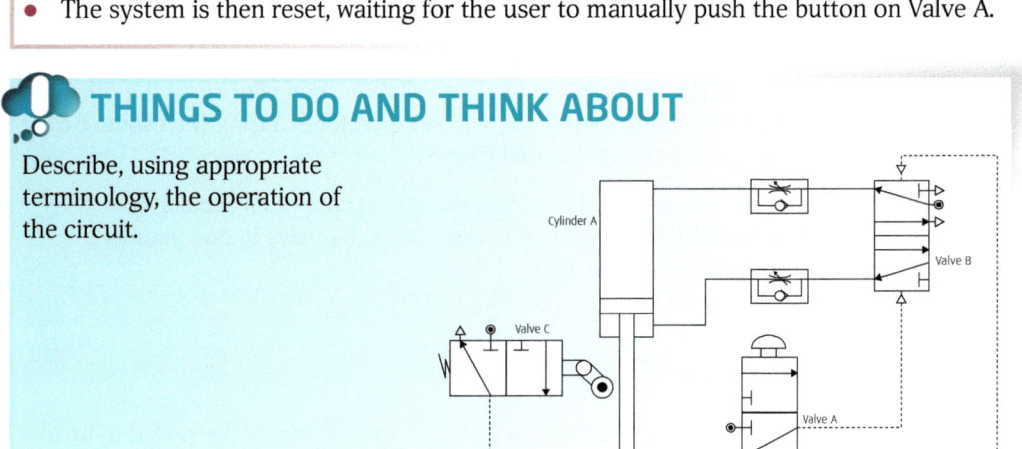

ONLINE TEST

Test your knowledge of this topic at www.brightredbooks.net

MECHANISMS AND STRUCTURES

PNEUMATIC SYSTEMS: AUTOMATIC CIRCUITS AND CASCADE SYSTEMS

AUTOMATIC CIRCUITS

Automatic circuits are how pneumatic circuits are typically utilised within industry. This is not only to help speed up production but also to allow for uniformity, making sure that everything is done in the exact same way to the exact same standard. There are two types of automatic circuits that are used.

Semi-Automatic Circuits

A semi-automatic circuit is one that has already been covered in this book. A semi-automatic circuit is one where a set process will complete once a human operator has started it.

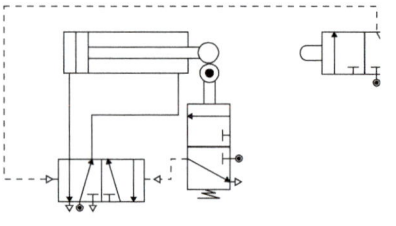

In this example, the circuit will only operate when an electronic signal has been sent into the solenoid-actuated spring-return value. In this case, it is a signal from the microcontroller. This will send pilot air to the 5/2 valve, causing it to change state. This causes the double acting cylinder to out-stroke. The cylinder, when out-stroked, will activate the plunger-actuated 3/2 valve. This will then change state and send pilot air back to the 5/2 valve, causing the cylinder to in-stroke. The process is then reset and ready to begin again once another electronic signal has been sent from the microcontroller.

Fully Automatic Circuits

A fully automatic circuit is one that will continue to work the sequence, performing the task over and over again, without any manual intervention. When mains air is supplied, the circuit will start and continue to work.

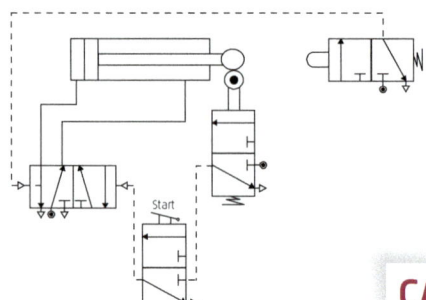

This circuit works by the double acting cylinder in-stroking. As this in-strokes, the roller on the 3/2 valve will actuate and send pilot air to the 5/2 valve, which will then also change state. This will then out-stroke the double acting cylinder, which will hit the plunger on the other 3/2 valve. This will then send pilot air to the other side of the 5/2 valve, changing its state again, and therefore causing the cylinder to in-stroke. This process will then continually repeat.

A fully automatic circuit can be interrupted, though, in case of an accident or emergency. This can be done by putting a lever–lever 3/2 valve in one pilot line. This can then be used as on/off switch.

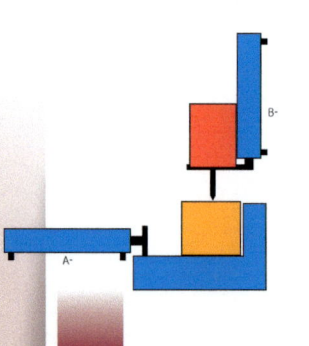

CASCADE SYSTEMS

Within a complex sequential control circuit, an extra 5/2 valve is sometimes added to break the circuit into separate parts. By 'grouping' areas of the circuit, it allows the cylinders to work independently of each other and not have to follow particular sequences.

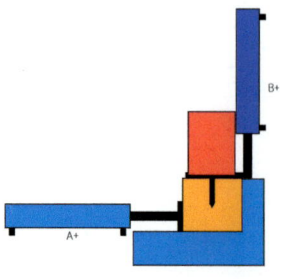

For example, a factory uses a pneumatic circuit by using out-stroking cylinders to hold material in place, and then to drill a hole in it. When the circuit starts, it is desired for the circuit to follow the sequence A+, B+, B−, A−.

This would not be possible, though, without an extra group-control 5/2 valve. This valve creates a cascade system that will prevent the old signal from remaining in the circuit to oppose any new signal. Essentially, it separates the circuit into two distinct parts – one where the circuit works with the left side

contd

Mechanisms and Structures: Pneumatic Systems: Automatic Circuits and Cascade Systems

of the group-control 5/2 valve, and one that will work with the right side of this valve. This will allow Cylinder B to out-stroke and then in-stroke, while Cylinder A remains out-stroked. This means that the air to valve 2 and valve 3 must be controlled separately to prevent them both from signalling Cylinder B's 5/2 valve at the same time and therefore locking it in its present state.

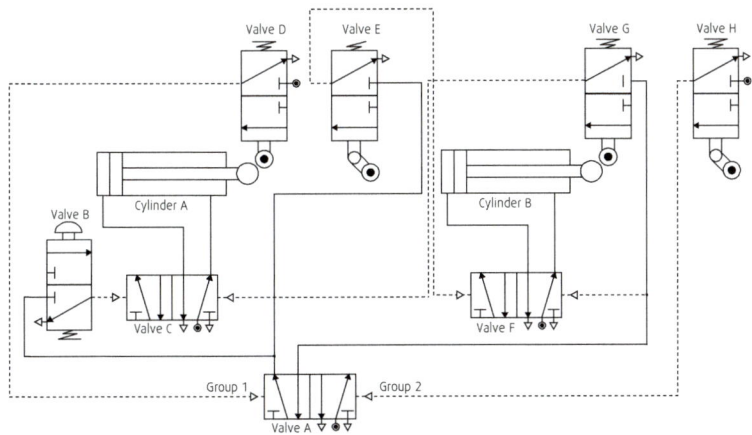

By building up a circuit with this extra group-control 5/2 valve, the circuit will now be possible.

- When Valve B is pressed, Valve A connects to Group 1, and it sends pilot air into 5/2 Valve C.
- This will actuate Valve C, causing it to change state and make Cylinder A out-stroke.
- The roller trip on Valve E then trips and sends pilot air to 5/2 Valve F, making Cylinder B out-stroke.
- As it out-strokes, it hits the roller trip on 3/2 Valve H.
- This sends pilot air to the 5/2 Valve A and changes the state of the 5/2 valve, and now supplies the circuit with Group 2 air.
- This will then send air to Valve F, causing Cylinder B to in-stroke.
- At the same time, this will also send air to Valve G. When Cylinder B in-strokes, it will actuate this valve and send pilot air to Valve C.
- This will cause the in-stroking of Cylinder A.
- Valve D then actuates, returning the supply to Group 1 air.
- The system is then reset, waiting on Valve B's push-button being pressed again.

THINGS TO DO AND THINK ABOUT

Describe, using appropriate terminology, the operation of the circuit shown.

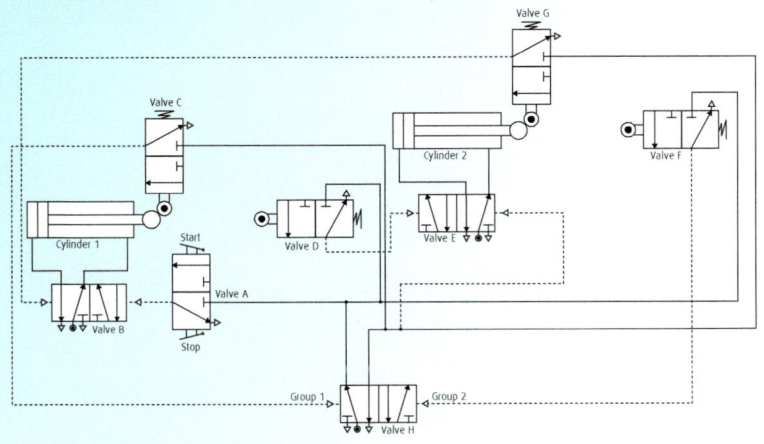

DON'T FORGET

It is a good idea to answer a question like this in bullet points. It will make it easier for you to answer, as you won't get lost as easily as you might when writing a paragraph.

DON'T FORGET

If you have to describe a group air circuit, **always** refer to the changing of groups.

DON'T FORGET

If you run out of space in the exam answering a question like this, there are spare pages at the back of the answer paper. Just make sure you write 'see on back' or something similar to make it clear to the marker that you have continued your answer somewhere else.

ONLINE TEST

Test your knowledge of this topic at www.brightredbooks.net

MECHANISMS AND STRUCTURES

STRUCTURES: MOMENTS AND REACTIONS

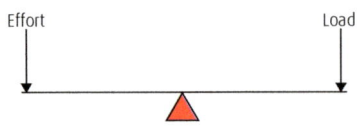

WHAT ARE MOMENTS AND REACTIONS?

The turning effect that occurs on a structure due to forces is called a **moment**.

When any system is in perfect balance like in this diagram, it is said to be in **equilibrium** or, better still, in **static equilibrium**. This means that not only is the structure in balance, but it is also stationary. That is what we have to calculate when dealing with any structure to ensure it does not collapse. In the case of moments, if it was not in equilibrium, the structure would fail by rotating either clockwise or anti-clockwise.

If the right-hand side of this system, in this case the load, has the higher force placed on it, it would rotate the beam clockwise. The contrary can be said for the other side – if the left-hand part of the system had the larger force, it would then cause the structure to turn anti-clockwise. Either way, the structure will collapse. For this beam to be in static equilibrium, the sum of all the clockwise and the sum of all the anti-clockwise must equal each other. In other words, the sum of the moments **must** equal 0.

$\Sigma_M = 0$

Clockwise moments = anti-clockwise moments

In National 5, you also learned that there are forces that will affect a structure on the vertical plane which are known as **reactions**. In this case, it has to be ensured that any forces pushing down on a system are equalled pushing back up.

Upward reactions = downward reactions

$\Sigma_R = 0$

Example:

By looking at this free-body diagram, calculate the reaction force R_2 by taking moments about R_1.

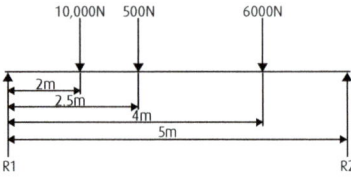

As it says in the question, moments must be taken about R_1. This essentially means that for the first part of our equation, R_1 is like a pivot point.

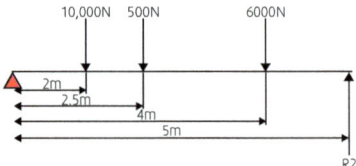

$\Sigma_M = 0$
$\Sigma CWM = \Sigma ACWM$
$(10{,}000 \text{ N} \times 2 \text{ m}) + (500 \text{ N} \times 2{\cdot}5 \text{ m}) + (6{,}000 \text{ N} \times 4 \text{ m}) = R_2 \times 5 \text{ m}$
$R_2 = (20{,}000 + 1{,}250 + 24{,}000) / 5$
$R_2 = \underline{\textbf{9{,}050 N}} \uparrow$

Now the value of R_2 is known, R_1 can be calculated by reactions.

$\Sigma_R = 0$
$\Sigma \text{ Up} = \Sigma \text{ Down}$
$R_1 + R_2 = 10{,}000 + 500 + 6{,}000$
$R_1 + 9{,}050 \text{ N} = 16{,}500$
$R_1 = \underline{\textbf{7{,}450 N}} \uparrow$

Therefore the reactions for the beam supports are

$R_1 = \underline{\textbf{7{,}450 N}}$ and $R_2 = \underline{\textbf{9{,}050 N}}$

UNIFORMLY DISTRIBUTED LOAD

A uniformly distributed load, sometimes referred to in its shortened form of UDL, is a load which is spread constantly along a length of the beam.

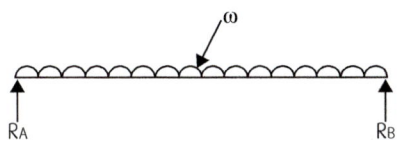

'ω' is the load acting along the length and is given in terms of the total force acting on each metre length of the beam. To work this out as a single force that can be used, the UDL is seen to be concentrated at its own centre of gravity. This means that the total force it will have is focused at the mid-point of its length. By multiplying ω by the total distance, this force can be found.

Example:

In this example, the 40 Nm distance is to be multiplied by the total distance it is acting over. In this case, it is 5 m.

UDL = 40 × 5
= **200 N**

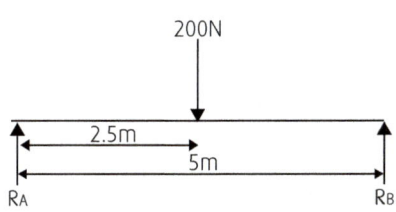

This gives a concentrated load of 200 N in the centre of the UDL length, so it is drawn as a single force of 200 N acting 2·5 m from the edge.

Now that this is known, the principle of moments can be used to calculate the reaction points R_A and R_B.

$\Sigma_M = 0$
$\Sigma CWM = \Sigma ACWM$
(200 N × 2·5 m) = R_2 × 5 m
R_2 = 500 / 5
R_2 = **100 N ↑**

$\Sigma_R = 0$
Σ Up = Σ Down
$R_1 + R_2 = 200$
$R_1 + 100 = 200$
R_1 = **100 N ↑**

 **DON'T FORGET**

This whole section of the course is based on the assumption you had a good grasp of it in National 5. If you didn't, read back through the 'Structures: Moments and Reactions' section of the National 5 Engineering Science Study Guide and use it to complete National 5 past paper questions from the SQA website.

THINGS TO DO AND THINK ABOUT

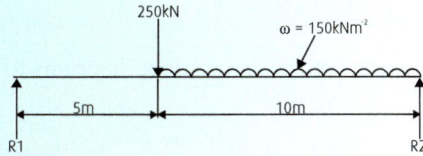

A simply supported beam supports a point load 5 m from the left edge, and a UDL of 150 kNm^{-2} over a 10 m length, as shown here.

a) By taking moments about R_2, calculate R_1.

b) Calculate R_2 using reactions.

c) Build and test this using a simulation package such as beamguru.com to ensure your answers are correct.

 **ONLINE TEST**

Test your knowledge of this topic at www.brightredbooks.net

MECHANISMS AND STRUCTURES

STRUCTURES: FORCES AT ANGLES

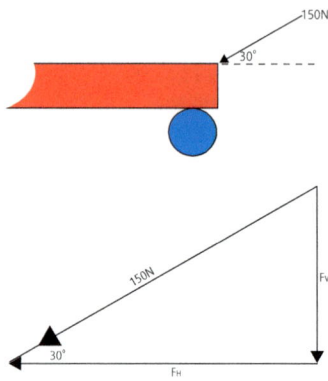

RESOLUTION OF A SINGLE ANGLED FORCE

In the real world, it is unlikely that forces will always be acting straight up or down. It is far more likely that any force acting on a structure will be acting at an angle.

Example:
In the diagram shown, you can see that the force is acting downwards at a 30° angle.

To calculate this force, it has to be split into two separate components – the **vertical** component, which shall be known as F_V, and the **horizontal** component, which shall be known as F_H.

To be able to do this, two things have to be known first – the force's **magnitude** and its **direction**. Once these are known, trigonometry can be used to solve this.

By redrawing the diagram and breaking it up into the separate components, it can be seen that the addition of F_V and F_H creates a triangle. This allows our known force to be the hypotenuse of this triangle, with the vertical component becoming the opposite, and the horizontal component being the adjacent.

By using SOH-CAH-TOA, the missing values can then be calculated. The horizontal value can be found by using cos $x°$ = adj/hyp, and the vertical force can be found using sin $x°$ = opp/hyp.

Horizontal force (F_H)
$\Sigma F_H = 0$
$\cos 30° = F_H / 150$ N
$F_H = 150 \times \cos 30°$
$F_H = 150 \times 0.866$
$F_H = \underline{129.9 \text{ N}}$ ←

Vertical force (F_V)
$\Sigma F_V = 0$
$\sin 30° = F_V / 150$ N
$F_V = 150 \times \sin 30°$
$F_V = 150 \times 0.5$
$F_V = \underline{75 \text{ N}} \downarrow$

DON'T FORGET
It is good practice to draw the direction of the force once you have discovered the amount. This can be done with a simple arrow after your answer.

MOMENTS AT ANGLES

Using the same principle, we can use this in moment diagrams to work out an unknown force.

Example 1:
In the diagram shown, a force of 500 N is hitting the beam at 45°, which will cause it to turn in a clockwise direction. At the other side from the pivot point, Force F is pulling down on it, causing anti-clockwise movement – but, for this structure to be in equilibrium, these two forces must equal each other.

In this case, both the horizontal and vertical forces do not need to be found. As the horizontal force would be pushing directly from the right, it would have no effect on the system as it creates no turning force. The vertical force **will** have an effect, though, as the downward force will try to turn the system in a clockwise direction.

To work this out, the vertical aspect of the 500 N force needs to be calculated:

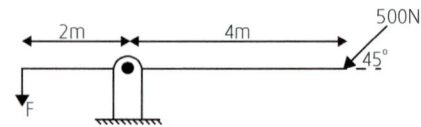

$\Sigma F_V = 0$
$\sin 45° = F_V / 500$ N
$F_V = 500 \times \sin 45°$
$F_V = 500 \times 0.707$
$F_V = \underline{353.6 \text{ N}} \downarrow$

ONLINE
Head to the BrightRED Digital Zone for a link on building and simulating different structures.

This figure can now be inserted into the moment's calculation:

$\Sigma CWM = \Sigma ACWM$
$(353.6 \text{ N} \times 4 \text{ m}) = (F \times 2 \text{ m})$
$1{,}414.4 = 2F$
$F = 1{,}414.4 / 2$
$F = \underline{707.2 \text{ N}} \downarrow$

contd

Mechanisms and Structures: Structures: Forces at Angles

Example 2:

If the beam is stepped, then both the horizontal and vertical aspects of the angled force need to be found, as clockwise moments will be produced by the vertical aspect of the 75 N load, and anti-clockwise moments will be produced by the horizontal aspects.

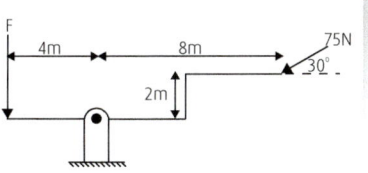

$\Sigma F_V = 0$
$\sin 30° = F_V / 75 \text{ N}$
$F_V = 75 \times \sin 30°$
$F_V = 75 \times 0·5$
$F_V = \underline{37·5 \text{ N}} \downarrow$

$\Sigma F_H = 0$
$\cos 30° = F_H / 75 \text{ N}$
$F_H = 75 \times \cos 30°$
$F_H = 75 \times 0·866$
$F_H = \underline{64·95 \text{ N}} \leftarrow$

$\Sigma_M = 0$
$\Sigma CWM = \Sigma ACWM$
$37·5 \text{ N} \times 8 \text{ m} = (F \times 4 \text{ m}) + (64·95 \text{ N} \times 2 \text{ m})$

You will notice here that the distance used for the horizontal aspect of the angle is multiplied by is 2 m instead of the expected 8 m. This is because when it is stepped like this, it is the height of the step that is causing the turning moments, and not the length it is over. Therefore it is the height of the step away from the pivot point that is used.

$300 = 4F + 129·9$
$4F = 300 - 129·9$
$4F = 107·1$
$F = 170·1 / 4$
F = 42·5 N

Example 3:

In this example, the angles are taken from the horizontal plane of the beam; but this is not always the case. Sometimes the angle that it shows will be from a vertical plane instead of the horizontal one. **Always** change this so that you are taking your angle from the horizontal.

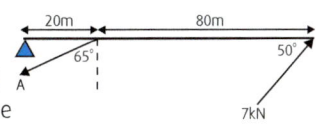

This will now allow you to always use the sin calculation, making things easier for yourself.

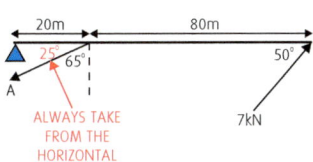

$\Sigma_M = 0$
$CWM = ACWM$
$(A \times \sin 25) \times 20 \text{ m} = (7 \text{ kN} \times \sin 50°) \times 100 \text{ m}$
$(A \times 0·423) \times 20 = (7 \times 0·766) \times 100$
$8·452 A = 536·231$
$A = 536·231 / 8·452$
A = 63·4 kN

> **DON'T FORGET**
> When doing any sort of moments, the distance **always** comes from the pivot point.

> **ONLINE**
> Follow the link at www.brightredbooks.net to build and simulate different structures.

> **ONLINE TEST**
> Test your knowledge of this topic at www.brightredbooks.net

THINGS TO DO AND THINK ABOUT

A crane with two lifting cables is used to lift bridge sections for a crossing over a motorway, and is represented by the free-body diagram shown.

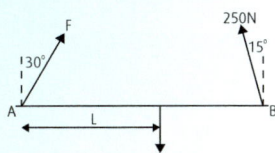

a) Calculate the force, F, in the cable acting at A, by resolving horizontally.

b) Calculate the weight, W, of the bridge section.

c) If the length of the bridge section between points A and B is 5 m, calculate the distance, L, by taking moments.

MECHANISMS AND STRUCTURES

STRUCTURES: RESULTANTS AND CONCURRENT FORCES

VIDEO LINK

Watch the video on the Digital Zone to see some examples of calculating the resultant.

DON'T FORGET

When dealing with horizontal forces, cos is to be used. If you are dealing with vertical forces, use sin.

DON'T FORGET

It is a good idea to draw an arrow showing the direction of the force once you have worked out the force and the angle. This makes the exam-marker aware that you understand how the forces are acting, and it could potentially be worth a mark.

THE RESULTANT

The **resultant** is the vector sum of two or more vectors. Basically, it is one single force that will have the same effect on an object as all of the forces combined. This is something engineers will sometimes have to discover when looking at a structure so that they know the overall force hitting an object, and which direction it will hit from. In National 5, you would have discovered this by drawing force diagrams, but in Higher this is more complex, and maths has to be used to work this out.

Example:

To work out the resultant, the overall horizontal and vertical forces need to be discovered, just as before. This time, though, a different approach to obtain the answer will be more beneficial – and this will include using negative numbers. If an arrow is going up when working out the vertical force, this is seen as a positive number. If the arrow is pointing downward, this is a negative number. The same approach is used for the horizontal forces – if the arrow is going to the right, it is a positive number, and if it is pointing to the left, it is a negative number. Think of it like creating a graph, and you are using the x and y axes.

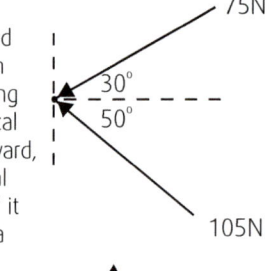

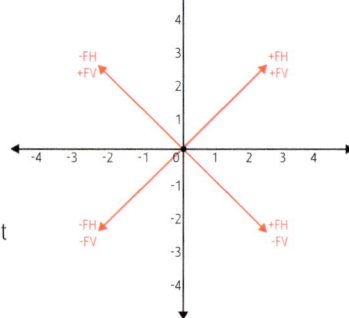

$\Sigma F_H = 0$
$F_H = (-75\cos30°) + (-105\cos50°)$
$F_H = (-75 \times 0.866) + (-105 \times 0.643)$
$F_H = -64.95 - 67.52$
$F_H = -132.47$ N

Don't worry if you obtain a negative number. All this means is that the force is pushing to the left. If you were calculating the verticals and your answer was negative, it would just mean that it is pushing downwards.

$\Sigma F_V = 0$
$F_V = (-75\sin30°) + (105\sin60°)$
$F_V = (-75 \times 0.5) + (105 \times 0.866)$
$F_V = -37.5 + 90.93$
$F_V = 53.43$ N

Once these two forces have been calculated, Pythagoras' Theorem can be used to work out the resultant.

Resultant² = $F_H^2 + F_V^2$
Resultant² = $(-132.47)^2 + (53.43 \text{ N})^2$
Resultant² = $17,548.3 + 2,854.8$
Resultant² = $20,403.1$
Resultant = $\sqrt{20,403.1}$
Resultant = 142.8 N

Once the resultant force has been gauged, the angle at which this force hits needs to be found. To do this, tangency is used.

$\varnothing° = \tan^{-1}(Rv / Rh)$
$\varnothing° = \tan^{-1}(53.43 / 132.47)$
$\varnothing° = \tan^{-1}(0.403)$
$\varnothing° = 24.4°$
This means the resultant = 142.8 N ↖ 24.4°

Mechanisms and Structures: Structures: Resultants and Concurrent Forces

CONCURRENT FORCES

Within a structure diagram, sometimes the concurrent force needs to be worked out. This is the single force that ensures static equilibrium is happening, as it opposes all other forces – for example, when a crane lifts an object.

By focusing on the hook, it can be seen that the two cables holding the object will be pulling it down. This means for the crane to hold the object and be in equilibrium, the lifting cable must contain the same force to stop this from dropping.

To calculate the value of R, we follow the same procedure as we would to work out the resultant.

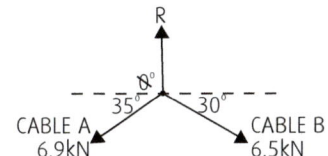

$\Sigma F_H = 0$
$R_H = (-6.9\cos35°) + (6.5\cos30°)$
$R_H = (-6.9 \times 0.819) + (6.5 \times 0.866)$
$R_H = -5.65 + 5.63$
$\mathbf{R_H = \underline{-0.02 \text{ kN}}}$

$\Sigma F_V = 0$
$R_V = (-6.9\sin35°) + (-6.5\sin30°)$
$R_V = (-6.9 \times 0.574) + (-6.5 \times 0.5)$
$R_V = -3.69 - 3.25$
$\mathbf{R_V = \underline{-6.94 \text{ kN}}}$

Resultant$^2 = F_H^2 + F_V^2$
Resultant$^2 = (-0.02)^2 + (6.94)^2$
Resultant$^2 = 0.0004 + 48.1636$
Resultant $= \sqrt{48.164}$
Resultant $= \underline{6.9 \text{ kN}}$

$\varnothing° = \tan^{-1}(Rv / Rh)$
$\varnothing° = \tan^{-1}(\underline{-6.94} / \underline{-0.02})$
$\varnothing° = \tan^{-1}(347)$
$\varnothing° = \underline{89.8°}$
This means $\underline{R = 6.9 \text{ kN} \nwarrow 89.8°}$

 DON'T FORGET

If you are unsure of what way the direction is facing, look at the diagram. In what directions are the largest forces going? Is it up or down? And is it left or right?

THINGS TO DO AND THINK ABOUT

Two boats are tied to the same mooring post at a dock. The forces acting concurrently on the mooring post are shown in the diagram.

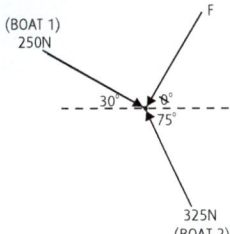

Calculate the magnitude of the reaction force R, and the angle θ.

 ONLINE TEST

Test your knowledge of this topic at www.brightredbooks.net

MECHANISMS AND STRUCTURES

STRUCTURES: COMPLEX RESOLUTIONS OF A FORCE

EXAMPLES AND SOLUTIONS

Sometimes, when calculating the different parts of a structure, the components are known. Instead, it is the force that is to be calculated. This makes it far more difficult, and may necessitate the use of simultaneous equations to work it out.

Example 1:

When approaching this type of question, break it down into its horizontal and vertical components to simplify it. By looking at the horizontal aspects first, the forces which are acting on it and pulling it to the left have to be equal to the forces that are pulling it to the right. The 35 kN would not be considered for this part of the solution – this is purely a vertical force and has nothing to consider when horizontal is being calculated.

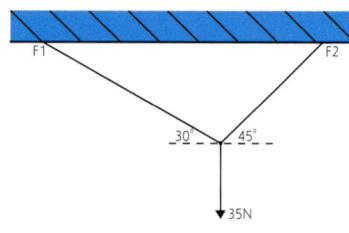

VIDEO LINK

Watch the video at www.brightredbooks.net to see a structural engineer going through a step-by-step guide to answer questions like these.

$\Sigma F_H = 0$
$\Sigma \leftarrow = \Sigma \rightarrow$
$F_1 \times \cos 30° = F_2 \times \cos 45°$
$F_1 \times 0·866 = F_2 \times 0·707$
$0·866 F_1 = 0·707 F_2$

$\Sigma F_V = 0$
$\Sigma \uparrow = \Sigma \downarrow$
$(F_1 \times \sin 30°) + (F_2 \times \sin 45°) = 35 \text{ kN}$
$0·5 F_1 + 0·707 F_2 = 35N \rightarrow 0·707 F_2 = 35 - 0·5 F_1$

Now we have these 2 statements, simultaneous equations can be used to resolve the missing force by transferring any relevant data from one equation into the other. As you can see, we now have 2 different equations where $0·707 F_2$ is a common denominator. We can see that both **$0·866 F_1$** equal this, but so does **$35 - 0·5 F_1$**. Therefore...

$0·866 F_1 = 35 - 0·5 F_1$
$1·366 F_1 = 35$
$F_1 = 35 / 1·366$ **$= 25·6N$ (3 S.F.)**

Now we know what F_1 is equal to, we can transfer this into one of the equations to discover F_2

$0·866 F_1 = 0·707 F_2$
$0·866 \times 25·6 = 0·707 F_2$
$22·1696 = 0·707 F_2$
$F_2 = 22·1696 / 0·707$ **$= 31·4N$ (3 S.F.)**

Example 2A:

In more complex diagrams, all the forces will be at angles, but this can still be approached in the same manner.

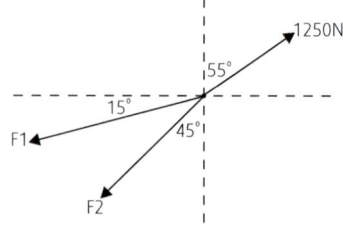

$\Sigma F_H = 0$
$\Sigma \leftarrow = \Sigma \rightarrow$
$(F_1 \times \cos 15°) + (F_2 \times \cos 45°) = (1250N \times \cos 35°)$
$0.966 F_1 + 0·707 F_2 = 1023·94$
$\Sigma F_V = 0$
$\Sigma \uparrow = \Sigma \downarrow$
$(1250N \times \sin 35°) = (F_1 \times \sin 15°) + (F_2 \times \sin 45°)$
$0·259 F_1 + 0·707 F_2 = 716·971$

The difficulty is that there is now two seperate calculations that don't easily fit into teach other like the previous example.

$0·966 F_1 + 0·707 F_2 = 1023·94$
$0·259 F_1 + 0·707 F_2 = 716·971$

contd

It can be noticed though that they do have something in common: $0.707F_2$. If the calculations are rearranged so they are foccusing purely on this figure, two different equations can now be discovered that equal $0.707F_2$. This must therefore mean that the answer to one of these equations has to equal the answer to the other.

$0.707F_2 = 1023.94 - 0.966F_1$
$0.707F_2 = 716.971 - 0.259F_1$
Therefore... $1023.94 - 0.966F_1 = 716.971 - 0.259F_1$

This can now be simplified to discover the value of F_1.

$1023.94 - 0.966F_1 = 716.971 - 0.289F_1$
$-0.966F_1 + 0.259F_1 = 716.971 - 1023.94$
$0.707F_1 = 306.969$
$F_1 = 306.969/0.707$
$F_1 = \underline{434N}$

Now F_1 is known, it can now be placed into one of the equations to calculate the value of F_2.

$0.966F_1 + 0.707F_2 = 1023.94$
$(0.966 \times 434) + 0.707F_2 = 1023.94$
$419.341 + 0.707F_2 = 1023.94$
$0.707F_2 = 1023.94 - 419.341$
$F_2 = 604.4094/0.707$
$F_2 = \underline{855N}$

Example 2B:

Another approach to a complex system such as this one, is to rotate it. Rotating it does not change any the forces, but it does allow for the removal of one of the angled forces. This rotation would allow one of the forces to become purely vertical or purely horizontal, hence removing the need for simultaneous equations.

$\Sigma F_H = 0$
$\Sigma \leftarrow = \Sigma \rightarrow$
$(F_1 \times \cos 60°) = (1250N \times \cos 80°)$
$(F_1 \times 0.5) = (1250N \times 0.174)$
$0.5 F_1 = 217.5$
$F_1 = \underline{434N}$

$\Sigma F_V = 0$
$\Sigma \uparrow = \Sigma \downarrow$
$(1250N \times \sin 80°) = (F_1 \times \sin 60°) + F_2$
$1231.1 = 0.866 F_1 + F_2$
$1231.1 = (0.866 \times 434) + F_2$
$1231.1 = 375.844 + F_2$
$F_2 = \underline{855N}$

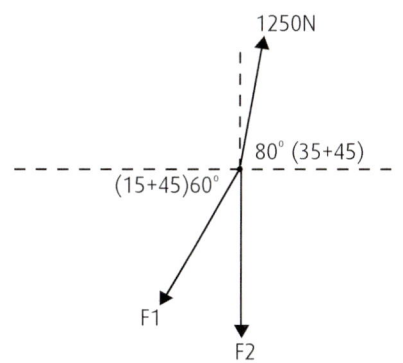

> **+ DON'T FORGET**
>
> It may say in a question within the question paper "solve by simultaneous equations", but this doesn't necessarily mean this is what you have to do. This is mentioned to give candidates a hint if they get confused if they cannot get a 'proper' answer. You have to answer things in the way that makes sense to you. If you get the correct answer, you cannot be wrong.

> **▶ VIDEO LINK**
>
> Go to the BrightRED Digital Zone to see a structural engineer answer a question like this, but with symmetry involved.

THINGS TO DO AND THINK ABOUT

A structural engineer has been asked to design the structure for a new tram-station roof in Edinburgh. If the node is in static equilibrium, calculate the magnitude of the forces in members F_1 and F_2.

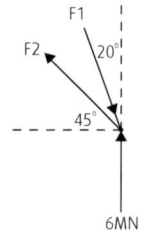

> **✓ ONLINE TEST**
>
> Test your knowledge of this topic at www.brightredbooks.net

75

MECHANISMS AND STRUCTURES

STRUCTURES: HINGED SUPPORTS

HINGE AND ROLLER SUPPORTS

Hinge and roller supports are used within a structure when there is a possibility that the beam may move sideways. The hinge support is capable of resisting forces acting in any direction, but it does not provide any resistance to rotation. For example, a door: when you are pulling the door open, you are creating a horizontal force. The hinge changes this into a rotational force and allows the door to open.

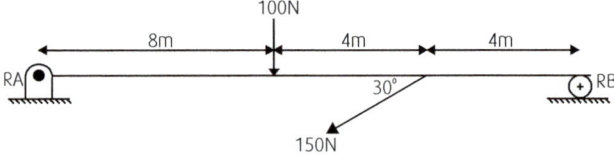

To calculate a hinged support like this, the reaction at a roller support is always at right angles to the surface.

When calculating R_A, the direction of it must always be assumed – at least to begin with. If any of the components work out as negative values when calculating the answer, this just means that the direction will be opposite to what was assumed. The reaction at the hinge support can be any angle and can work in any direction.

Within the diagram, there are three unknown quantities:

1. the magnitude of reaction R_B
2. the magnitude of the reaction R_A
3. the direction of the reaction R_A.

To calculate these unknowns, a lot of previous knowledge needs to be combined. Firstly, to calculate R_B, the knowledge of moments needs to be used.

$\Sigma_M = 0$
$\Sigma CWM = \Sigma ACWM$
$(100 \text{ N} \times 8 \text{ m}) + (150\sin30° \times 12 \text{ m}) = R_B \times 16 \text{ m}$
$800 + (150 \times 0.5 \times 12) = 16 R_B$
$800 + 900 = 16 R_B$
$16 R_B = 1,700$
$R_B = \underline{106.3 \text{ N}} \uparrow$

To work out R_A, our knowledge of resolving a resultant can be utilised. The hinge has not only an unknown force but also an unknown angle.

$\Sigma F_V = 0$
$\uparrow = \downarrow$
$RA_V + 106.3 = 100 + (150 \times \sin30°)$
$RA_V = 100 + (150 \times 0.5) - 106.3$
$RA_V + RB = 100 + 75 - 106.3$
$RA_V = \underline{68.7 \text{ N}}$

$\Sigma F_H = 0$
$\rightarrow = \leftarrow$
$RA_H = (150 \times \cos30°)$
$RA_H = (150 \times 0.866)$
$RA_H = \underline{-129.9 \text{ N}}$

Resultant$^2 = F_H^2 + F_V^2$
Resultant$^2 = 129.9^2 + 68.7^2$
Resultant$^2 = 16,874 + 4,719.7$
Resultant $= \sqrt{21,593.7}$
Resultant $= \underline{146.9 \text{ N}}$

$\emptyset° = \tan^{-1}(Rv / Rh)$
$\emptyset° = \tan^{-1}(68.7 / 129.9)$
$\emptyset° = \tan^{-1}(0.529)$
$\emptyset° = \underline{27.9°}$
$R_A = \underline{146.9 \text{ N}} \nearrow 27.9°$
$R_B = \underline{106.3 \text{ N}} \uparrow$

LOADED SUPPORTS

On many occasions extra supports are through triangulation to give the structure extra strength and to ensure it can support the needed forces acting upon it.

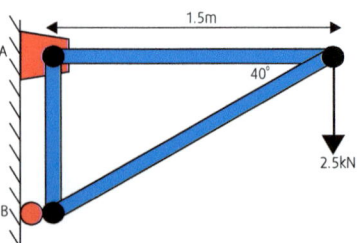

Within a structure such as this, a lot of the force is being pushed into B, so a structural engineer would have to work out the magnitude and direction of B to ensure stability. To do that you must first use the SOH-CAH-TOA calculation from trigonometry to calculate the length of AB.

> **DON'T FORGET**
>
> This isn't a huge step in difficulty from the things you have already learned. If you are struggling with this, go back to the 'easier' stuff and make sure you have a strong grasp of it. This essentially is just moments and reactions, but adding in a concurrent force where the hinge is.

contd

Mechanisms and Structures: Structures: Hinged Supports

Tan = opp / adj
tan40° = AB / 1·5
AB = 1·5 tan40° = **1·259m**

Once this has been done, the moments calculation can be used to work out the forces around the hinge A.

CWM = ACWM
2·5 × 1·5 = 1·259 × RB
RB = 3·75 / 1·259 = **2·98kN**

(this will only act in a horizontal manner as it only has one directional component, like the way a roller in a hinge and roller system)

To calculate the forces acting upon A, we treat this also in the same manner as a normal hinge and roller support. We know the horizontal aspects of this system as it must equal the forces at B, so we now need to discover the vertical ones.

$\Sigma F_V = 0$
↑ = ↓

AF_V = **2·5 kN** ↑

Resultant² = $F_H^2 + F_V^2$
Resultant² = 2·98² + 2·5²
Resultant² = 8·3521 + 6·25
Resultant = √14·6021 = **3·82kN**

$\varnothing° = \tan^{-1}(R_v / R_h)$
$\varnothing° = \tan^{-1}(2·5 / 2·98)$
$\varnothing° = \tan^{-1}(0·94) = $ **43·2°**

R_A = **3·82 kN** ↗ **43·2°**
R_B = **2·98 kN**

VIDEO LINK

Follow the link at www.brightredbooks.net to simulate structures such as these. The same link can be used as a calculator, so after you have worked it out yourself, use it to mark your answer.

THINGS TO DO AND THINK ABOUT

A TV is attached to the wall using a wall bracket. The image shown displays the forces acting on the bracket. (For clarity, the bracket has been rotated to the horizontal plane.)

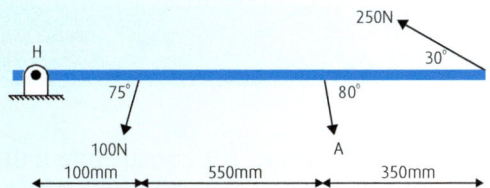

a) Calculate the force, A, exerted by the wall anchor.

b) Calculate the **magnitude** and **direction** of the reaction force at the hinge, H.

ONLINE TEST

Test your knowledge of this topic at www.brightredbooks.net

MECHANISMS AND STRUCTURES

STRUCTURES: FRAMED STRUCTURES

USES OF FRAMED STRUCTURES

A framed structure is the most common way of construction in the modern world. This is due to their ability to allow for large, strong yet light structures to be built. Framed structures are built by joining different beams, known as members, together by a joint called a node. This is also usually done in a collection of triangular shapes. Making a framed structure by triangles is a technique called **triangulation** and is used to provide strength and support. When a force is applied to a triangular frame, two of its members stretch the third one, making it tense. This in turn pulls the other two members towards it, making the structure rigid and spreading the force across all three members of the triangle.

Framed structures using this method can be seen every day, from cranes and bridges to electricity pylons.

A member in a framed structure can be classified as one of two things – either a **tie** or a **strut** – and this is dependent on what type of force it is under. When resolving issues in a frame structure, the **magnitude** and **nature** of the forces in each of the members need to be determined. This means that, in addition to the size of the force in the member, you will be expected to state whether the member is a **strut** or a **tie**.

A **strut** is a member that is under a compressive force. This means the member is being squashed.

A **tie** is a member that is under tensile force. This means the member is being pulled apart.

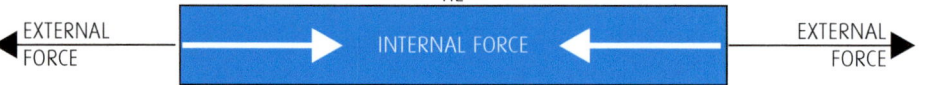

When working out the forces on a structure, it is also important that you are able to realise the nature of the forces acting on it. That means you have to understand whether each member is a strut or a tie. To do this, consider how the force is acting on it.

With the flowerpot pulling down on the frame, the upper member can be seen to be in **tension**, as it is being **pulled**. That means this member is a **tie**.

The diagonal beam, though, will be being **compressed**, as it is being **pushed** back towards the wall. That means this member is a **strut**.

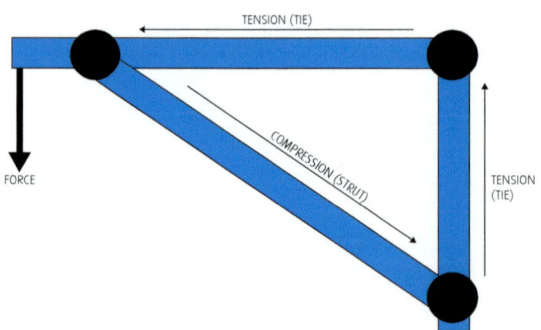

Mechanisms and Structures: Structures: Framed Structures

The same technique is also used for larger structures. By working out the nature of each member, it helps an engineer to understand how the forces act upon a structure, which in turn will help to ensure its stability.

This example shows a gantry used for hanging different traffic lights to control traffic. As the weight of the traffic lights pulls down on the structure, it affects each member in different ways.

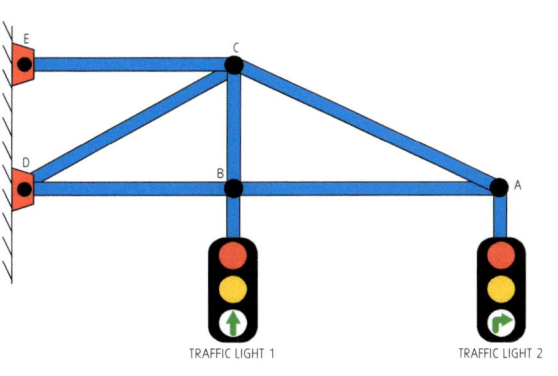

- As traffic light number 2 hangs from the structure, it can be seen that it is stretching member AC, meaning that **AC** must be a **tie**.
- AB, though, is being pushed towards the wall by the same force pulling down on traffic light 2. As it cannot move left because of the wall, it is being compressed. Therefore **AB** is a **strut**.
- BC is being pulled down by traffic light 1 – and, as it cannot move without the structure failing, **BC** must be a **tie**.
- BD is also being pulled down by the weight of traffic light 1. As this also cannot move without the structure failing, **BD** must be a **tie**.
- The same can be said for CE. It is being affected by traffic light 1 in a pulling motion, but held in place at Node E. This means **CE** must be a **tie**.
- CD though, will, be under compression between the forces acting at D and C when traffic light 1's weight is pulling down on the frame. Therefore **CD** is a **strut**.

DON'T FORGET

You don't have to remember and state the name of each type members, all you have to do is realise the nature of the forces acting on it - this means if it's easier you can refer to the member being in compression or tension rather than being a strut or tie.

VIDEO LINK

Watch the video on the BrightRED Digital Zone to obtain a greater understanding of the forces acting upon a frame structure.

ONLINE TEST

Test your knowledge of this topic at www.brightredbooks.net

THINGS TO DO AND THINK ABOUT

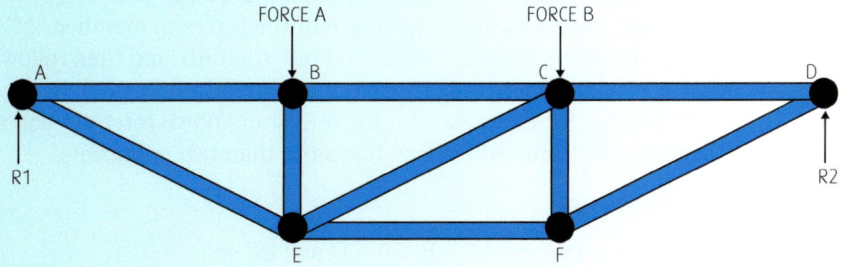

This free-body diagram shows the structure used to lift items on a skyscraper construction site from one floor to another. Force A and Force B represent the forces acting on the structure from the lift motor and cable drum. What is the nature of each of the members?

MECHANISMS AND STRUCTURES

STRUCTURES: NODAL ANALYSIS 1

WHAT IS NODAL ANALYSIS?

To calculate the forces acting on each node within a structure, we use a technique called **nodal analysis**. Nodal analysis is based around the fact that if a structure is in equilibrium, then each node must also be in equilibrium. This therefore means that the sum of the forces acting on any of the nodes must equal zero.

When solving frame-structure questions, some analysis of the structure is required to determine the best starting point and which of the conditions of static equilibrium to apply first.

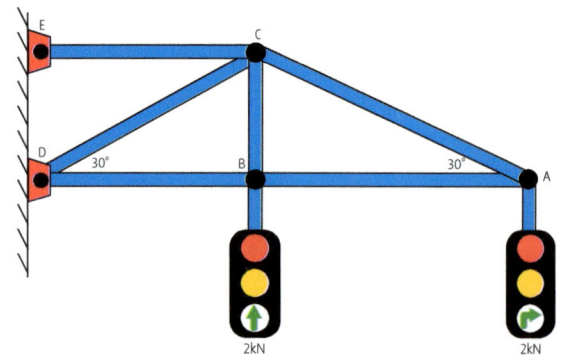

By looking at the frame, and how the forces are acting upon it, it can be determined whether each member is a strut or a tie. This will then allow us to understand how the internal forces on each member are acting, and which way the arrow will be pointing towards the node.

AC = Tie

AB = Strut

CB = Tie

DB = Tie

EC = Tie

CD = Strut

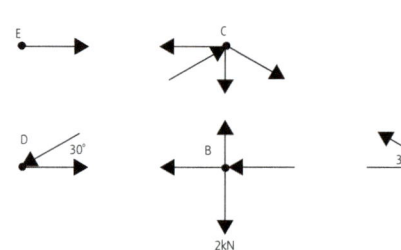

To calculate the forces acting on each node, they have to be taken one node at a time. By working out the horizontal and vertical forces acting on each node, it will give this information as well as giving the figures needed to then work out the forces acting upon the other nodes.

To discover which node to start with, there are two approaches. Firstly, look at the question. If it asks 'Calculate the magnitude and nature of the forces in members AC, AB, CB, CD and CE', then AC should be the first one to find out, then AB, and then follow in this fashion. The other approach should only be used if the question does not state where to start – and that is by looking for the node with the fewest unknown forces. If you are taking this approach, **never** start somewhere that has more than two unknowns.

Working Out Your Response

In this example, the forces in members AC, AB, CB, CD and CE still need to be discovered. As stated above, since the question asks for AC first, this should be the starting point. It is ALWAYS good practice to then redraw this individual node and all the information about it that is known. This should help to avoid confusion and prevent you from becoming overwhelmed, as it is now just a single diagram to resolve, instead of the larger and more intimidating and complex one.

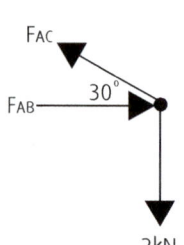

From this diagram, the vertical force can be calculated, which will allow the value of AC to be determined. Then the horizontal forces can also be calculated, and this will allow the value of AB to be discovered.

$\Sigma F_V = 0$
$F_{AC}\sin 30° - 2 = 0$
$0.5 F_{AC} = 2$
$\mathbf{F_{AC}} = \underline{\mathbf{4 \text{ kN (tie)}}}$

$\Sigma F_H = 0$
$F_{AB} - 4\cos 30° = 0$
$F_{AB} = 4 \times 0.866$
$\mathbf{F_{AB}} = \underline{\mathbf{3.5 \text{ kN (strut)}}}$

> **DON'T FORGET**
>
> STRUT: ←INTERNAL FORCE→
> TIE: →INTERNAL FORCE←

Now that the forces in these members are known, another node can be approached. The question now asks you to work out CB, which means that two nodes could be used – either node C or node B. Node C has many unknown quantities, whereas node B has fewer. This means that node B should be the next one approached.

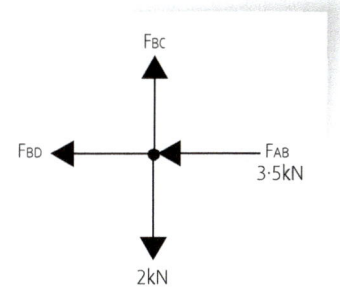

$\Sigma F_V = 0$
$F_{CB} - 2 = 0$
$\mathbf{F_{CB} = \underline{2 \text{ kN (tie)}}}$

This now reduces the unknown forces on node C, which allows further analysis of it and the discovery of the forces acting upon it.

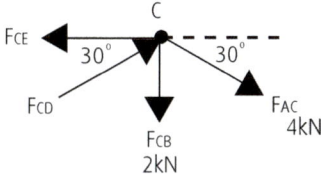

$\Sigma F_V = 0$
$F_{CD}\sin 30° - 2 + 4\sin 30° = 0$
$0.5 F_{CD} = 2 + 2$
$\mathbf{F_{CD} = \underline{8 \text{ kN (strut)}}}$

$\Sigma F_H = 0$
$-F_{CE} + F_{CD}\cos 30° + 4\cos 30° = 0$
$F_{CE} = (8 \times 0.866) + (4 \times 0.866)$
$\mathbf{F_{CE} = \underline{10.4 \text{ kN (tie)}}}$

VIDEO LINK

Watch the video at www.brightredbooks.net to see a worked example of a nodal-analysis question like this.

THINGS TO DO AND THINK ABOUT

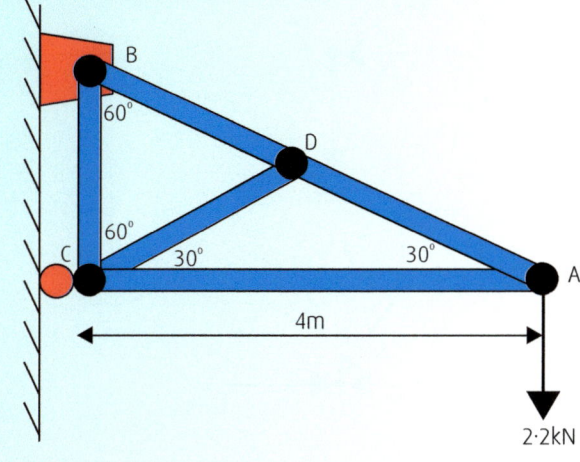

ONLINE TEST

Test your knowledge of this topic at www.brightredbooks.net

A frame structure that holds a shop sign is shown here.

a) Calculate the reaction force at C, using the principle of moments.

b) Calculate the reaction force at B.

c) Calculate, using nodal analysis, the magnitude and nature of the forces in members AD and AC.

MECHANISMS AND STRUCTURES

STRUCTURES: NODAL ANALYSIS 2

MORE ON NODAL ANALYSIS

When analysing the nodes of a structure, sometimes it is not as easy as it first seems, and previous mathematical knowledge needs to be used. In this example, a frame for supporting a bungee-jumping platform is suspended from the top of a building.

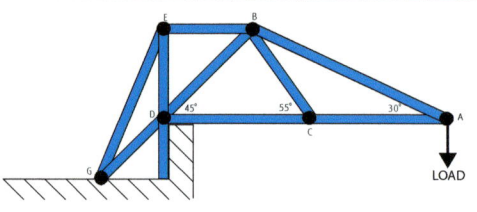

The magnitude and nature of the forces in members AB, AC, BC, CD and BD can be calculated for a load of 7 kN acting on the support frame, using nodal analysis.

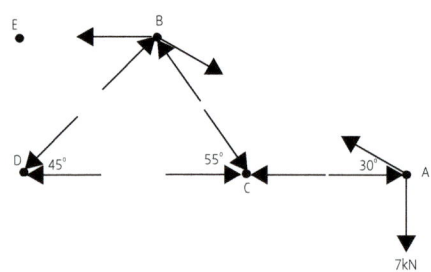

As with previous examples, it is worthwhile taking the time to figure out whether each member is a tie or a strut first, and therefore how the internal forces will act on each member. This can be drawn directly onto the diagram, or separately, but it is worthwhile doing now, as it will help to ensure that the arrows are facing in the correct directions in the node diagrams.

As with the previous example, it has to be approached one node at a time. By working out the horizontal and vertical forces on each node, it will show the magnitude of each member's force, as well as giving us relevant data to work out the forces on the other nodes.

Node A:

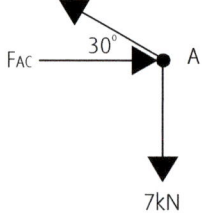

$\Sigma F_V = 0$
$F_{AB}\sin 30° - 7 = 0$
$0{\cdot}5\,F_{AB} = 7$
$F_{AB} =$ 14 kN (tie)

$\Sigma F_H = 0$
$F_{AC} - F_{AB}\cos 30° = 0$
$F_{AC} = 14 \times 0{\cdot}866$
$F_{AC} =$ 12·1 kN (strut)

> **DON'T FORGET**
>
> Start on the node it asks you to do first. In this case, AB needs to be found first.

Node C:

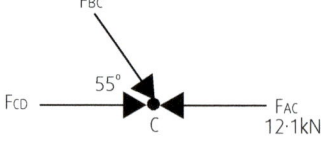

$\Sigma F_V = 0$
$-F_{BC}\sin 55° = 0$
$F_{BC} =$ 0 kN

As this member has been proven to have no force in it, this means it is a **redundant member**. This just means it is a member in the structure that has no forces going through it, and it is not there necessarily for support.

$\Sigma F_H = 0$
$F_{CD} - F_{AC} = 0$
$F_{CD} =$ 12·1 kN (strut)

Node B:

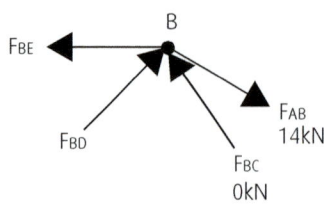

With this diagram, you can see that there are several unknown angles. This means we have to analyse other parts of the structure, and use our knowledge of angles and triangles to discover these.

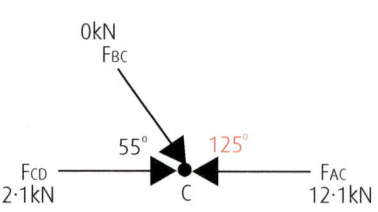

82

contd

By looking again at node C, it can be seen that F_{BC} is hitting at 55°. As a straight line is 180°, that means the angle between F_{BC} and F_{AC} **must** be 125°.

Knowing that the angles of a triangle add up to 180° also helps us to discover the angle between F_{BD} and F_{BC}. This means this angle must be 80°.

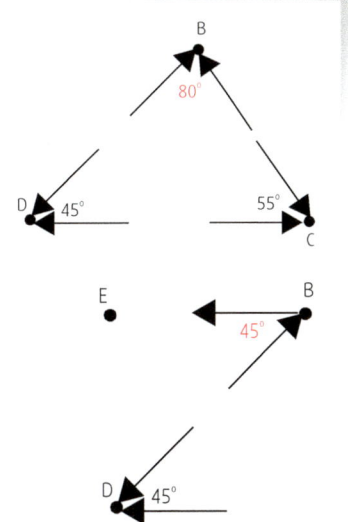

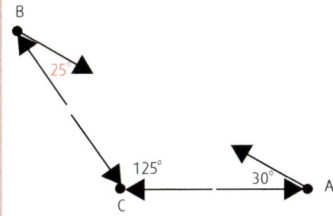

Using the same principle, the angle between F_{BC} and F_{AB} must be 25°.

Using the theory of Z angles, it can also learned that the angle between F_{BE} and F_{BD} is 45°.

And finally, it can be revealed that F_{AB} is pulling down at 30°, since it is known that all the angles in a straight line must add up to 180°.

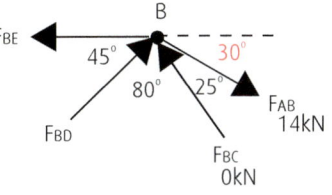

Now that this is all known, it will allow the discovery of the final force, F_{BD}.

$\Sigma F_V = 0$
$F_{BD}\sin 45° + 0 - F_{AB}\sin 30° = 0$
$0.707\, F_{BD} = 14 \times 0.5$
$F_{BD} = 7 / 0.707$
$\mathbf{F_{BD} = \underline{9.9\ kN\ (strut)}}$

VIDEO LINK

Watch the video on the BrightRED Digital Zone to see an engineer work through a nodal-analysis question like this to work out how to resolve the forces.

💭 THINGS TO DO AND THINK ABOUT

This diagram shows a structure supporting a hopper used for pouring concrete into mixer trucks.

For a particular load, members DF and EG are both subjected to a tensile force of 90 kN.

a) Using nodal analysis, calculate the **magnitude** and **nature** of the forces in members CE, DE, CD, BD and BC.

b) Two strain gauges attached to member AC are also shown, and are used to indirectly monitor the weight of concrete in the hopper.

Explain why the resistance of strain gauge RG2 **increases** as the load in the hopper increases, whereas the resistance of strain gauge RG1 does **not** change as the load increases.

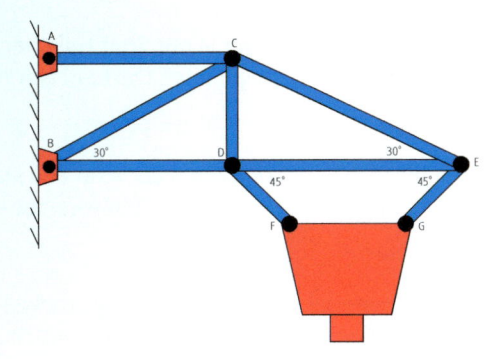

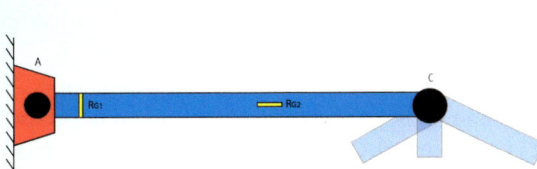

ONLINE TEST

Test your knowledge of this topic at www.brightredbooks.net

MECHANISMS AND STRUCTURES
MATERIALS: MATERIAL PROPERTIES

PROPERTIES AND BEHAVIOURS OF MATERIAL

A structural engineer's expertise lies not only in the design of a structure but also in the understanding of the properties and behaviour of the material from which it is made when put under different conditions. It is not only the shape of a structure, or how the members are joined, that influences a structure's overall performance, but also the material that it is made from. If any single member in a frame structure was to fail, it would create a domino effect on the other members, and ultimately the structure would collapse. So, what it is made from has to be fit for purpose. In order to select the correct material, the structural engineer must research and consider a range of different types of materials, or modify materials in specific ways to improve the performance, durability and cost-effectiveness of the structure.

When doing this, the properties that the engineer would consider are:

Strength

The strength of a material is its ability to resist a force. All materials have some degree of strength, but the greater the force that the material can resist, the stronger it is.

Some materials, like mild steel, can be strong in tension but weak in compression. The opposite is also true for some materials, like concrete, which are strong in compression but weak in tension. This is the reason some materials are mixed to create a material that holds multiple properties. An example of this would be reinforced concrete, which is concrete reinforced with mild steel. This allows it to have the compressive strength of concrete as well as the tensile strength of steel.

Elasticity

The elasticity of a material is its ability to return to its original shape or length once a load or force acting on it has been removed. A material such as rubber is described as having high elasticity, as it can be stretched but will return to its original condition once the force acting on it stops.

Plasticity

A material with plasticity is one that will change its shape or length under a load, but will stay this deformed shape once the load is removed. Clay has high plasticity.

Ductility

A ductile material is one that can bend and stretch without breaking, allowing it to be formed into thin sheets or very thin wire, for example. Copper is seen as a ductile material.

Brittleness

A brittle material is one that is easily cracked, snapped or broken when a force is applied. It is the opposite of a ductile material, which also means it has very little plasticity. A brittle material, such as glass, will fail under loading without stretching.

Malleability

If a material is malleable, it means it can be shaped, worked or formed without fracturing. It is a property that is very closely related to plasticity. An example of a malleable material is gold, where it can be changed into multiple objects – from jewellery to being flattened into a metal leaf without breaking.

It is very likely that a material will have multiple properties, and in the modern world they are frequently coated or mixed with another material to ensure that particular material properties are achieved. For example, steel is strong, but it corrodes easily due to its iron content. This can be changed, though, by covering it in particular paints.

DON'T FORGET

It is worthwhile making sure you understand these material properties and how they would affect a member, as they link strongly within the rest of this topic.

Mechanisms and Structures: Materials: Material Properties

MATERIALS TESTING

In order to discover the properties of a material, a sample must undergo some form of material testing. There are many different types of test available, but the most commonly used one is the tensile test by using a tensometer. As the name suggests, the material is subjected to a tensile force by stretching it and pulling it apart. The results from a tensile test allow the following properties to be determined:

1. the elasticity of a material
2. the plasticity or ductility of the material
3. the ultimate tensile strength of the material.

The results of a typical tensile test for a sample of mild steel are shown in the graph.

The shape of the graph is very important and is used to help engineers predict how materials will behave or react under different forces or loading conditions.

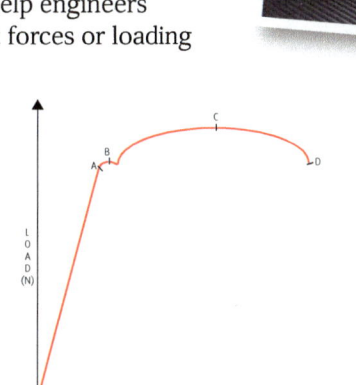

- **Between points 0 and A**, the material will behave elastically and is known as the **elastic region**. This means that the material will stretch under the load in this area, but it will return to its original length when this load is removed.

- **Point A** is known as the **limit of elasticity**. Any loading beyond this point results in a plastic deformation of the sample.

- **Point B** is called the **yield point**, and a permanent change in length will result when this is reached, even when the load is removed. Loading beyond this point will result in a rapidly increasing extension.

- The area **between points B and D** is known as the **plastic range**. This is so called because between these two points on the graph the material will behave in a plastic manner.

- **Point C** is known as the **ultimate load**. This is the point where the maximum load or force that the material can withstand is reached.

- **Between points C and D**, the amount of force the material can now withstand deteriorates rapidly, and the sample will start to 'neck'. This means the cross-sectional area of the specimen is starting to become thinner as it is pulled further apart.

- The sample eventually reaches **point D**, where it breaks or fractures. This is known as the **breaking point**.

 VIDEO LINK

Go to www.brightredbooks.net to see a specimen material being tensile tested.

 VIDEO LINK

Go to www.brightredbooks.net to see how a tensometer works.

THINGS TO DO AND THINK ABOUT

The graph shown was produced using data from a tensile test on a specimen of material.

Redraw this on paper, and identify the following points on the graph.

- Yield point
- Elastic range
- Ultimate load
- Breaking point
- Plastic range.

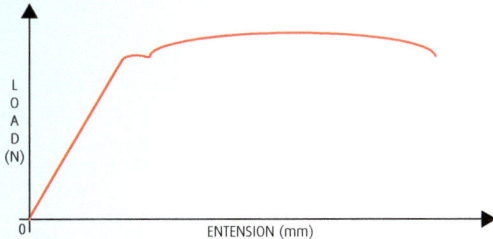

 ONLINE TEST

Test your knowledge of this topic at www.brightredbooks.net

85

MECHANISMS AND STRUCTURES

MATERIALS: STRESS AND STRAIN

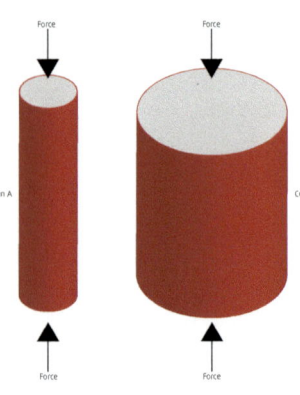

STRESS

As you know from your previous studies in National 5, **stress** occurs when a direct force or load is applied to a member of a structure. The effect that occurs on this particular member will depend on the cross-sectional area of it.

In the example, you can see that Column B has a greater cross-sectional area than column A. If the exact same load was applied to each column, then column A will be under greater stress.

Stress is calculated using the formula:

$\sigma = F / A$

(Stress = Force ÷ Area).

- **Force** is measured in **Newtons (N)**
- **Area** is measured in **square metres (m²)**
- The unit for measuring **stress** is **Newtons per square metre (Nm⁻²)**

Example:

A wire that is 5 mm in diameter is subjected to a 200 N force. Calculate the stress that the wire is put under.

$\sigma = F/A$
$= 200 / (3.14 \times 2.5^2)$
$= 200 / 19.625$
$= \underline{\mathbf{10.2\ Nmm^{-2}}}$
$= \underline{\mathbf{0.01\ Nm^{-2}}}$

DON'T FORGET

If you have to find the area of a circle, both πr^2 and $\pi d^2/4$ will find the answer. Use whichever one you are used to working with.

STRAIN

Strain is the response of a material when stress is applied. When the stress is applied to a member of a structure through a direct force or load, it will result in some form of deformation – usually a change in its length. This deformation is known as the **strain**. If the load that is applied to the structural member is compressive in nature, the length will therefore reduce. If a tensile load is applied, then the length will instead increase. Every material changes its shape to some extent when put under stress, although this may not always be visible to the human eye.

Strain is calculated using the formula:

$\varepsilon = \Delta l / l$

(Strain = Change in Length ÷ Original Length).

- **Length** is measured in **metres (m)**
- **Strain** <u>does not</u> have a unit, as the lengths in both parts of the equation cancel each other out.

Example:

A 4-metre-long steel cable is used to support a load. When the load is applied, the wire is found to stretch by 3.5 mm. Calculate the strain on the wire.

$\varepsilon = \Delta l / l$
$= 3.5 / 4,000$
$= \underline{\mathbf{0.0009}}$

STRESS-STRAIN GRAPHS

When the stress and strain of a material are discovered, a **stress–strain graph** can be created to display the findings. At first glance, this looks very similar to a load-extension graph, but you will notice the axis labels are different, suggesting that it is something else. When the material data is displayed in this form, it makes it easier for engineers to interpret the information more easily in design situations.

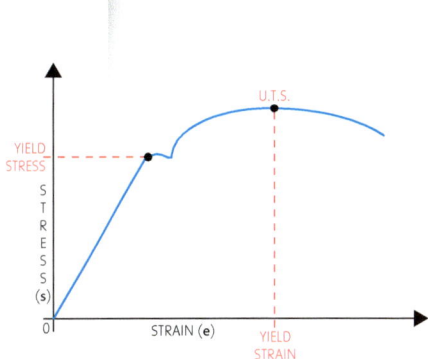

Looking at the graph, several things can be learned. First is the **yield stress**, i.e. the maximum stress that can be applied to the specimen without causing a permanent change in its length. In any safe structure, the loading that is to be applied should never produce a stress that is greater than the yield stress. The **yield stress point** is at the end of the elastic range, which means that the member should therefore remain elastic under loading.

The **ultimate tensile stress** can also be found out using this graph. The ultimate tensile stress, commonly known as UTS for short, is the maximum stress the material can withstand before it starts to fail. If a member in a structure is loaded beyond this point, it will start to stretch, the material's cross-section will therefore reduce, and the member will then quickly fail.

VIDEO LINK

Head to the Digital Zone to watch a video explaining how material properties can be discovered using stress-strain graphs.

The **yield strain** can also be discovered using this graph. This is the maximum percentage of plastic extension produced in the specimen before it starts to fail under loading. By taking the UTS as a reference point, the yield strain can be easily found. This needs to be known for a ductile material like copper so it can be formed into pipes or wires. For this to happen, it needs to be stretched to a point where it will no longer go back to its original shape, or deteriorate due to being stretched too much.

Type of Material

Using these graphs, it can also be worked out what each type of material is.

For example, you can see that material A has very little strain for the high amount of stress it can withstand. This means that although it is very strong, it must also be brittle. This means its fracture will be sudden with little to no plastic deformation. This material is likely to be something like glass or cast iron.

Material B is strong material that is not very ductile – it has a plastic region, but this is small. This means it will stretch, but not by a huge amount before it breaks suddenly.

Material C is a ductile material, as its strain is relative to the amount of stress it is put under. The material has a large elastic range, and the plastic range is clearly visible in comparison to the other samples. The ultimate load is also noticeable, and this is where it will start to deform and deteriorate before it hits the breaking point.

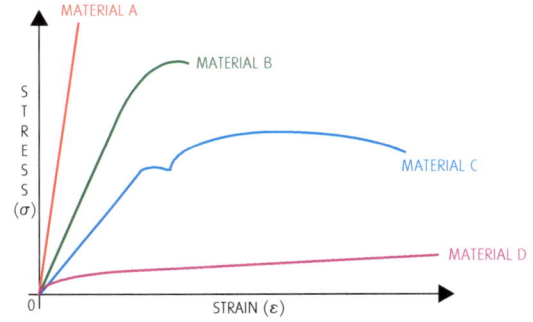

Material D has a very small elastic region but a large plastic region. It cannot take a huge amount of stress, but has a large strain. This means it will deform when a small load is placed on it, and it must have a high plasticity.

THINGS TO DO AND THINK ABOUT

A specimen was tested in a materials laboratory for potential use in a new structure. The results are given in the graph shown here.

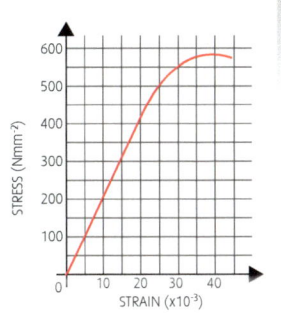

a) Describe the effects of applying a stress of 400 Nmm^{-2} to the specimen and then removing the stress.

b) Describe the effect of applying a stress greater than 550 Nmm^{-2} to the specimen.

ONLINE TEST

Test your knowledge of this topic at www.brightredbooks.net

MECHANISMS AND STRUCTURES
MATERIALS: YOUNG'S MODULUS

WHAT IS YOUNG'S MODULUS?

When any material is loaded past its elastic limit, its performance will deteriorate and it will become unpredictable. This could be potentially disastrous, or even fatal, when you consider what this structure might be. For this reason, when designing structures, engineers must ensure that all the expected stresses put on any member are held within the elastic range.

When looking at a material's stress–strain graph, the elastic limit is always a straight line. This means that stress is proportional to strain – and, knowing this, **Young's Modulus** can be calculated. Young's Modulus is sometimes called the **modulus of elasticity** and is calculated using the following formula:

E = σ / ε

(Young's Modulus = Stress ÷ Strain).

Example 1:

An aluminium beam is 1·7 m long and has a square cross-section of 25 mm × 25 mm. When a tensile load of 4·3 kN is applied, it produces a change in length of 0·5 mm. Calculate Young's Modulus for the beam.

σ = F/A = 4,300 / (25 × 25)
= **6·9 Nmm^{-2}**

ε = Δl /l = 0·5 / 1,700
= **0·3 × 10^{-3}**

E = σ /ε = 6·9 / 0·3 × 10^{-3}
= **23 kNmm^{-2}**

Example 2:

The second way that the information can be found out is by using the stress–strain graphs. Using the straight line of the elastic region, information can be gained to help you calculate the data required. This can be done anywhere that has known figures for both the stress and the strain.

In this stress–strain graph, it can be seen that the line hits the grid exactly at 25 Nmm^{-2} for stress, and its corresponding 5 × 10^{-3} strain. As these are both exact and known figures, this is a good place to work out the Young's Modulus.

E = σ / ε = 25 / 5 × 10^{-3}
= **5 kNmm^{-2}**

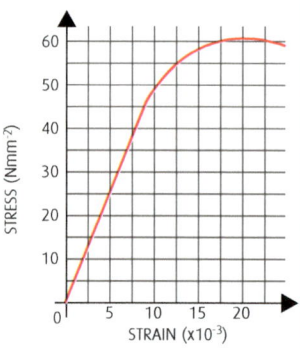

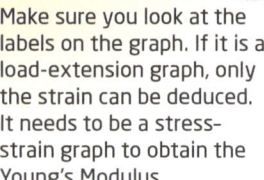

DON'T FORGET

Make sure you are using **exact** figures from the graph. When working out the Young's Modulus, there should never be a degree of estimation. That is how a structure fails, as you don't have exact data.

DON'T FORGET

Make sure you look at the labels on the graph. If it is a load-extension graph, only the strain can be deduced. It needs to be a stress-strain graph to obtain the Young's Modulus.

USING TABULAR DATA

When designing a structure, an engineer would not be expected to test every single piece of material that is used. The most commonly used materials will previously have been tested exhaustively, so instead the engineer would refer to the test data tables that are available in British Standards publications. An extract of these is available in your data booklet for use in your coursework and exams.

Material	Young's Modulus (E) kNmm^{-2}	Yield stress (σ) kNmm^{-2}	Ultimate tensile stress Nmm^{-2}	Ultimate compressive stress Nmm^{-2}
Mild steel	196	220	430	430
Stainless steels	190–200	286–500	760–1,280	460–540
Low-alloy steels	200–207	500–1,980	680–2,400	680–2,400

contd

Cast iron	120	–	120–160	600–900
Aluminium alloy	70	250	300	300
Titanium alloy	110	950	1,000	1,000
Nickel alloys	130–234	200–1,600	400–2,000	400–2,000
Concrete	–	–	–	60
Concrete (steel-reinforced)	45–50	–	–	100
Concrete (post-stressed)	–	–	–	100
Plastic (ABS polycarbonate)	2·6	55	60	85
Plastic (polypropylene)	0·9	19–36	33–36	70
Wood (parallel to grain)	9–16	–	55–100	6–16
Wood (perpendicular to grain)	0·6–1·0	–	–	2–6

An engineer would examine and use this information to help work out what material should be used for any given job. This table could also be used to give information about a material if one has already been chosen for the task. The table you have access to obviously doesn't show every single possible material available, but it has the most commonly used ones.

Example:
A supporting member for an extruding shop sign is in tension. When it is still in its elastic range and under a load of 700 N, it extends by 0·11 mm. The support wire has a diameter of 10 mm and a length of 2,500 mm.

a) Calculate Young's Modulus for the support wire.

b) State the name of the material used in the support wire.

Example:
a) Using the information given, Young's Modulus can be calculated:

$A = \pi r^2 = 3·14 \times 5^2$
= **78·5 mm²**

$\sigma = F / A = 700 / 78·5$
= **8·9 Nmm⁻²**

$\varepsilon = \Delta l / l = 0·11 / 2,500$
= **0·000044**

$E = \sigma / \varepsilon = 8·9 / 0·000044$
= **202 kNmm⁻²**

b) Now that the Young's Modulus has been discovered, the material used can be identified. By looking at the Young's Modulus column on the table, the only material that fits into our calculated range is a **low-alloy steel**, as its Young's Modulus is between 200 and 207.

DON'T FORGET
This table can be found in your data booklet, making it readily available for any assessment.

ONLINE TEST
Test your knowledge of this topic at www.brightredbooks.net

DON'T FORGET
Within Higher, it will rarely be as simple as using one calculation. It is more likely you will have to find out information by using other calculations before you can answer the question.

THINGS TO DO AND THINK ABOUT

A load-extension graph for a standard specimen is shown.

The test specimen was initially 100 mm long and had a cross-sectional area of 30 mm².

a) Calculate Young's Modulus for the test specimen.

b) State the name of the material.

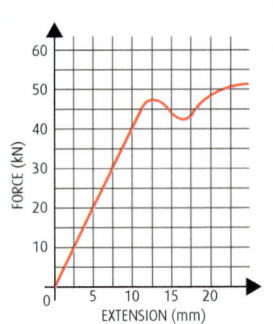

MECHANISMS AND STRUCTURES
MATERIALS: FACTOR OF SAFETY

REASONS FOR STRUCTURAL FAILURE

Although most structures are extremely safe and well designed, unforeseen circumstances can cause them to collapse. A structural engineer can never be absolutely certain that they have accounted for every single possible type of load or force that will affect the structure. So, when a structure fails, an investigation has to take place to discover the reason for this failure.

The most common reasons for a structural failure are:

Overloading

This is when the load or forces put upon a structure exceeds the value that was used during the design process. This type of failure may be due to the structure being used inappropriately, for example a fully grown adult riding a child's bike. It could also happen because the circumstances of its use have changed since the original design, for example a bridge that was designed and constructed before traffic was as heavy as it is in the modern world.

Another sudden change in loading on a structure to consider is the weather. The weather can be unpredictable in its nature, and nobody can predict with 100% certainty when heavy snow, rain or wind will happen, or when we will have extreme winds. Freak weather conditions can produce additional forces on a structure over and above what may have been calculated for a normal use.

Material Failure

The material of a structure may fail for several reasons. The material may not be of a consistent quality and may therefore have built-in flaws. It is also impossible to guarantee the performance of natural materials such as wood, as they inevitably contain natural defects such as knots.

Another reason for a structure to fail due to its material is natural deterioration or corrosion. Some materials are vulnerable to specific conditions, for example wood, which will swell up as it absorbs moisture. This is not always due to it being in constant contact with water; but swelling and warping can be caused just by unprotected wood being in contact with moisture through the air, condensation, or through contact with rain. When it then goes inside to a heat source, this moisture is removed, causing it to change shape. Metals will also corrode, and this needs to be considered. For example, copper will turn green, and any metal that contains iron, such as steel, will rust due to it unavoidably being in contact with oxygen and water.

Joint Failure

The joints or attachment methods used within the structure may fail if they are inappropriate or not strong enough to support the load. The structure may also fail if the joints have been inadequately crafted – for example poor welding. The welds on large structures have a huge force placed on them, and any imperfections will cause them to snap.

Fatigue

Fatigue is the weakening of a material due to repeated loading and unloading. This wears down the material's resistance to breaking, and eventually causes it to fail. If the loads are above a certain threshold, microscopic cracks will begin to form, and eventually a crack will reach a critical size. This crack will increase suddenly, and the structure will then fracture.

To show how fatigue can affect a structure, bend a paperclip forwards and backwards. A paper clip is designed to be bent. Even if it is bent beyond its expected distance, it

DON'T FORGET

If asked about this in the exam, make sure you are writing about the corrosion of a metal, and **not** rust. Rust is only one form of corrosion, and only happens to metals that contain iron.

contd

Mechanisms and Structures: Materials: Factor of Safety

probably still won't break first time. Eventually, though, it will snap due to the frequent movement – and, although it may not have an increased force applied to it, it will still fail.

Because of this, a **factor of safety** must be applied to the design when a risk analysis takes place. This factory of safety will vary from one structure to another, depending on the possible consequences of failure. For example, the factor of safety applied to a balance beam in a gymnasium will have a far lower factor of safety than that of the Queensferry Crossing road bridge. If the balance beam breaks, the gymnast will fall and may hurt themselves. If the Queensferry Crossing collapses, though, it would have catastrophic consequences, probably with multiple fatalities.

The factor of safety applied to a structure depends on two things. First is the safe working load/stress. This will be somewhere within the elastic range, and definitely not beyond the yield point. This ensures that if high loading is applied, it will not exceed the yield stress, and will therefore prevent permanent deformation. You may have noticed, within an elevator, a sign showing the weight limit or amount of people it will safely hold. Once this has been reached, or even passed, the elevator will not move. This is known as the **safe working load** (SWL).

Secondly, the ultimate load/stress needs to be considered. This is the point where the material will start to break, and its loading ability will deteriorate.

Factor of Safety = Ultimate load / Safe working load

Factor of Safety = Ultimate stress / Safe working stress

Example 1:

A mobile lifting hoist is used to help elderly and disabled people get into and out of bed.

Member BC is made from aluminium alloy, and when a particular force is put on the handle, it stretches it 0·7 mm. Calculate the factor of safety.

As we know, the calculation for the factor of safety is the **ultimate stress** divided by the **safe working stress**, so this needs to be discovered first.

The **ultimate stress** can be found by looking at the UTS column in the material data table within the data booklet. This states that the **UTS = 300 Nmm^{-2}**.

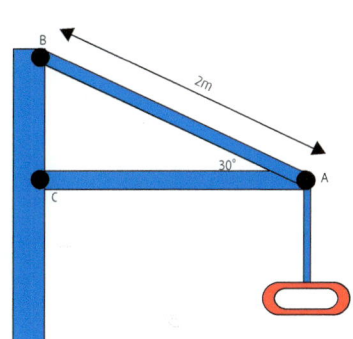

The **safe working stress** will need a little more work to determine. As you know, stress is calculated by dividing the force by the area – but neither of these details is given here. The only other calculation known that contains stress is the calculation for Young's Modulus. With a little transposition of the formulae, the stress can then be calculated. To do this, though, the Young's Modulus and the strain would need to be found first.

The Young's Modulus of the material can easily be found by once again looking at the material data table within the data booklet. This states that the Young's Modulus of aluminium alloy is **70 kNmm^{-2}**.

The strain can then be calculated using the information given.

$\varepsilon = \Delta l / l = 0{\cdot}7 / 2{,}000$
   = **0·00035**

Using the Young's Modulus calculation, the stress (and hence the safe working stress) can be found.

$E = \sigma / \varepsilon$
$70{,}000 = \sigma / 0{\cdot}00035$
$\sigma = 70{,}000 \times 0{\cdot}00035$
   = **24·5 Nmm^{-2}**

Now that the UTS and the SWS are known, the factor of safety can be determined.

F.o.S. = UTS / SWS = 300 / 24·5
   = **12·2**

THINGS TO DO AND THINK ABOUT

To help with the process of discovering the factor of safety, an engineer may have to ask themselves several questions about the structure to ensure that the F.o.S. is high enough. Name at least three things that would have to be considered.

VIDEO LINK

Go to the BrightRED Digital Zone to see a chair being fatigue-tested.

DON'T FORGET

Remember always to convert calculations to the same unit.

ONLINE TEST

Test your knowledge of this topic at www.brightredbooks.net

COURSE ASSESSMENT

OVERVIEW

Within this course, you will be assessed by two different means – by a question paper and by an assignment. Both of these are externally marked, with the assignment being completed in class and then sent off to the SQA for marking. Each means of assessment is carefully designed to give the opportunity to show your breadth and depth of knowledge, understanding and skills within different engineering contexts and challenges. Your knowledge and skills will be used to solve appropriately challenging practical engineering problems, within both practical and theoretical contexts.

ONLINE

Don't allow yourself to get too stressed. Yes, it is important to study – but also plan some short study breaks to take your mind off it for a period of time. Go to www.brightredbooks.net for some good advice on how to relax and de-stress yourself.

VIDEO LINK

Head to www.brightredbooks.net and watch the video on how to create an effective study plan.

DON'T FORGET

Remember to use your data booklet in any assessment within the course. It has all of the calculations and units that you will need within it.

ONLINE

Past papers, as well as the marking schemes, can be found on the SQA website. Use these within your studying to prepare yourself for the question paper, as well as the assignment, by completing questions and then marking them yourself. It will help you to realise not only what could be asked but also where the marks come from. Head to the BrightRED Digital Zone for a direct link!

EXAM

The question paper has a total mark allocation of 110. This is 69% of the overall marks for the course assessment, and it is designed to give the opportunity to demonstrate the skills, knowledge and understanding you have acquired and practised in relation to the following areas:

Area	Range of marks
The systems approach, energy and efficiency, engineering roles and disciplines, and the impacts of engineering	10–17
Analogue electronic control systems	20–35
Digital electronic control systems	15–25
Drive systems and pneumatics	10–20
Structures and forces	15–25
Materials	8–14

The question paper will consist of two separate sections.

Section 1

Section 1 is out of 20 marks, and will consist of short-answer questions. These are similar to the questions you have encountered throughout this book, where it will focus on one specific area of the course. By reading this book and completing these questions, along with your work in school, you will develop and better prepare yourself for them.

Section 2

Section 2 is out of 90 marks, and will consist of extended-response and structured questions. These may contain numerous parts of the course within one question, and are designed to show how many aspects of engineering overlap and how they can all be connected in a real-world situation. This section tests not only your knowledge but also whether you have the understanding of the course to comprehend and tackle this.

ASSIGNMENT

The assignment is designed to assess your ability to apply the skills and knowledge you acquired and developed during the course by putting them within a real-world scenario. This will involve designing and building/simulating potential solutions, as well as testing and evaluating existing ones.

Like the National 5 assignment, this is to be completed within an 8-hour period and is strictly closed-book. This means that you will not have access to any learning and teaching materials, the internet, exemplar materials or resources during this time. Your teacher also cannot give you any advice during the assignment, as they are clearly informed that no matter how you are progressing, they cannot interfere or give guidance.

contd

Course Assessment: Overview

Analysis

Within this area of assessment, you are likely to be asked about three things – drawing system diagrams, drawing sub-system diagrams, or to write/complete a specification.

Area	Range of marks
Analysis	4–8
Designing a solution	8–12
Building/simulating a solution	8–12
Testing	8–14
Evaluation	8–14

If you are asked to draw a **system diagram**, make sure you consider **all** possible inputs and outputs to the system. Take care to properly read the design brief to ensure you do this. The purpose of the **sub-system diagram** is to now break this system diagram down into all its possible components. This needs to be detailed and once again complete with all components to get full marks. You will have to show the interaction between the different sub-systems, which means you have to consider all parts of the system and all the components involved.

A **specification** should be a bullet-point list of the things the design must and must not do. This information will be found within the brief, so read through this with a fine-tooth comb and pick out all the information given.

Designing a Solution

Whether this be mechanical, structural or electronic, ensure you include any calculations to prove your design would work. It may also be worthwhile to include annotations next to your potential solution so that you can properly explain what you are thinking.

If you have to design any aspect of a solution, it should contain **detailed** bsi standard drawings of any gearing/mechanical system, with annotation explaining your solution, the number of teeth the gears have, the input and output to the system, the speed it will turn, and so on, to prove it will do what it is expected to do.

Building/Simulating a Solution

Whether this is mechanical, structural or electronic, ensure you are creating something that matches the brief or specification. Completing this section can be done by physically building the systems through equipment such as Fischer Technik or electronic breadboarding, or it can be done through suitable computer simulation software, such as Yenka. It does not matter which approach you choose – your marks will not differ. Once you have constructed/simulated any solution, ensure you have a clear print-out, and annotate it. Point out what is what on your design, and fully explain it.

If it asks you to build/simulate something that you can see is wrong, then build it exactly the way it shows you, as this may link with the testing aspect of the assessment, which would then involve pointing out the incorrect areas, and rebuilding.

Testing

Within this section, you are expected to have a logical and thorough explanation of any planned tests, a description of the expected results and the actual results achieved. If it asks you to write or complete a test plan or test table, ensure you are writing about **all** possible and potential tests to ensure that the outcome is correct. It is extremely important that you go through **every** single scenario you can, to test how the system works.
This should be logical and detailed – and you should make sure you are explaining and justifying all your actions and the appropriateness of the tests.

Evaluation

There are many ways of approaching an evaluation task, but whatever way you approach it, make sure you refer to **all** specifications, and prove how you know your design fulfils it.

One way you could approach this is to copy down each individual specification point and underneath (possibly in a different colour or style) write a statement saying 'I met this specification by …', and then prove how this has been fulfilled, and **exactly** how you know this.

It may also be useful to suggest how this solution could be further improved. To obtain full marks for this section, your evaluation needs to be clear, detailed and well argued. Ask yourself: what problems occurred, and how did you resolve them?

DON'T FORGET

If completing this on the computer, make sure you are frequently saving your work in case the machine crashes. It is a good idea to set up a specific folder for any work you create, and to make sure filenames are clear and suitable.

ONLINE

By looking at the SQA page for Higher Engineering Science, you will see that it gives you more details on the course and its assessment, as well as an exemplar for the assignment.

THINGS TO DO AND THINK ABOUT

When you receive your assignment, have a good read through it to ensure you understand what is needed. It is also worthwhile going through the marking criteria and to try to mark your own work when a section is complete. Make sure the work you are handing in covers everything asked, and you are attaining as many marks as you are capable of. Your teacher will not give you back your assignment to clear things up, or to further improve it.

APPENDICES

GLOSSARY

Amplifier: Part of an electronic circuit that can be found after an input device, and that is used to increase the signal.

Bearing: A part used within a gearing system that is designed to wear down. This allows for only the bearing needing to be replaced and extending the mechanical system's life.

Boolean expression: The equation that shows how a logic circuit works.

Clutch: A type of coupling that allows two rotating shafts to be connected and disconnected.

Compressive force: When something is being squashed. It is put under compression.

Concurrent force: The single force that ensures static equilibrium is happening. It opposes all other combined forces.

Coupling: A method of joining two different shafts together within a gearing system.

Darlington pair: Two transistors connected together to allow for the powering of large-output devices.

Driver: Part of an electronic circuit that will increase the power for a high-powered output device.

Engineering impact: The effects an engineering solution will have. This is in three sections – Social (the effect it has on society and people), Environmental (the effect it has on the ecosystem, wildlife and the planet) and Economic (the effect it has in relation to money).

Equilibrium: When a structure is in perfect balance.

Factor of safety: A number applied to a structure depending on the possible consequences of failure. This number depends on two things – the safe working load/stress and the ultimate load/stress.

Flowchart: A pictorial diagram showing how a microcontroller code works.

Group air: When a pneumatic circuit uses an extra 5/2 valve to allow for the separation of a circuit into two different ones. This is done to allow for cascading systems and for two different valves to be separately controlled.

Infrastructure: The basic systems and services that a region, country or organisation needs in order to work effectively. These include transport links and power supplies.

Knowledge: Refers to learning. Anything that can be read by a person, whether it is concepts, principles or information regarding a particular subject(s), is knowledge. This can be gained through books, media, encyclopaedias, academic institutions and many other sources.

Lubrication: A method of reducing friction within a system by a liquid such as oil or grease.

Magnitude: The amount of force acting upon an object.

Mark time: The amount of time a motor is on for during pulse-width modulation.

Member: A beam within a frame structure.

Microcontroller: The 'brain' of a programmable electronic circuit. The code is programmed into this to allow the circuit to complete set tasks.

Nodal analysis: A technique for framed structures, in which nodes are analysed to discover the nature and magnitude of each member.

Node: A joint within a frame structure.

Operational amplifier: An analogue processing device used to amplify a signal. These can be configured in many different ways to do different tasks.

Pulse-width modulation: A way of controlling the speed of a motor by quickly switching it on and off.

Redundant member: A member in a frame structure that has no forces going through it, and is not there necessarily for support.

Resultant: A single force that will have the same effect on an object as all of the forces combined.

Sequential control: When a system, in particular a pneumatic system, follows a particular sequence.

Shafts: rods used for putting gears onto within a gearing system.

Significant figures: The amount of numbers that your answer has to be given in. The first significant figure of a number is the first digit which is not zero. Hence the first significant figure of 2,009 is 2, and the first significant figure of 0·002009 is 2. Both numbers have 4 sig figs.

Skill: Skill refers to the ability of using knowledge and applying it in a context. Everyone can learn knowledge, but not everyone has the skill to use it.

Soft start: The technique of starting a motor slowly instead of starting at full speed.

Space time: The amount of time a motor is off for during pulse-width modulation.

Static equilibrium: When a structure is not only in balance but is also stationary.

Strut: A member in a frame structure that is under compressive force.

Tensile force: When something is being pulled apart. It is put under tension.

Tie: A member in a frame structure that is under tensile force.

Truth table: A table consisting of 1s and 0s showing how a logic diagram works.

Voltage-divider: An input part of an electrical circuit made up from two different forms of resistor to change real-world conditions, such as temperature or light, into a voltage.

INDEX

3/2 valves 60, 61
5/2 valves 61

amplification 26, 94
analogue control systems 10–11
analogue electronics 14–29
AND gate 30, 31, 37, 38, 39
Arduino 46–9
  counters 48–9
  'if' command 48
  writing and understanding code 46–7
assignment 4–5, 92–3
automatic circuits 66
automatic doors 49

ball bearings 57
base-emitter junction 26
bearings 56–7, 94
belt drive 53
bipolar junction transistor (BJT) 26
Boolean expressions 31, 94
  combinational 32–3
  creating circuit diagrams from 36–7
  discovering from brief 33
  discovering from circuit 32
breaking point 85
brittleness 84
bungee-jumping platform 82–3
bush/bushing 56

cascade systems 66–7
central heating systems 9, 24–5
chain drive 53
chemical engineering 6
circuit board tests 51
circuit diagrams, creating from Boolean expressions 36–7
civil engineering 6
clay 84
closed-loop control 8–9
  two-state 10
clutches 55, 94
code
  writing with Arduino 46–9
  writing with PBASIC 42–5
combinational logic 32–3
  truth tables 34–6, 94
comparator 10, 18–19
compound gear systems 52
compressive forces 78–9, 94
concrete 84
concurrent forces 75, 94
continuous loops 40
copper 84
counters 44–5, 48–9
couplings 54–5, 94
  flexible 55
  rigid 54
course assessment 4–5, 92–3
course content 4
crane 71, 75, 78
cylinders 60–1

Darlington pair 94
decision box 40
difference amplifier 10, 19, 24
digital electronics 30–9
'digitalRead' command 48
double inversion 39
drain 28, 29
drive systems 52–3
driven 52
driver 52, 94
ductility 84

economic impacts 7
efficiency 13
elastic region 95
elastic strain energy 13
elasticity 84
electric drill 55
electrical energy 13

electrical engineering 6
electronic circuits 14–15
electronic engineering 6
elevator 41
emerging technologies 7
energy 12–13
engineering
  disciplines 6
  impacts 7, 94
  roles 6
environmental engineering 6
environmental impacts 7
EOR gate 31, 35, 38
equilibrium 68, 94
  static 68, 75, 80, 94
exam 92
extended voltage-dividers 16

factor of safety 91, 94
fatigue 90–1
feedback, types 10–11
feedback signal 10
flange couplings 54
flat belt 53
flexi couplings 55
flowcharting 40–1, 51, 94
forces
  at angles 70–1
  complex resolutions of 72–3
  compressive 78–9, 94
  concurrent 75, 94
  direction 70
  magnitude 70, 94
  resolution of single angled 70
  resultants 74, 94
  tensile 78–9, 94
framed structures 78–9
friction 56–7
fully automatic circuits 66

gantry 79
gate 28, 29
gearing systems 52–3
glass 84
glossary 94
gold 84
group air 67, 94

hairdryer 8
heat energy 13
high-powered control 50
high signal state 42
hinge and roller supports 76
hopper 83

idler gear 52
'if ... else' command 44, 48
infrastructure 6, 7, 94
input resistor 20
inverting amplifier 20–1

joint failure 90
journal bearings 57

kinetic energy 13
knowledge 94
  skills versus 7

LDR graph 17
LED bulbs 12
limit of elasticity 85
loaded supports 76–7
logic gates 30–1
loops, continuous 40
low signal state 42
lubrication 56, 94

magnitude (of force) 70, 94
malleability 84
mark time 51, 94
materials 84–91
  behaviours 84

factor of safety 91, 94
failure 90
properties 84
strain 86–7
stress 86, 87
testing 85
using tabular data 88–9
Young's Modulus 88–9
mechanical engineering 6
mechanical power 58–9
mechanical systems 52–9
  clutches 55, 94
  coupling methods 54–5
  drive systems 52–3
  friction 56–7
  mechanical power 58–9
  torque 58
members 78–9, 94
microcontrollers 22, 42–50, 66, 94
modulus of elasticity 88–9
moments 68
  at angles 70–1
mooring post 75
MOSFETs 28–9
  as drivers 29
  in high-powered control 50
  saturation point 28–9
motor control 50–1
muff couplings 54

NAND equivalents 38–9
NAND gate 30, 31, 34
  as functionally complete 38
negative feedback 10, 20
nodal analysis 80–3, 94
nodes 78, 94
non-inverting amplifier 21
NOR gate 31, 35, 36, 38
NOT gate 30, 31, 38, 39
NPN transistor 26

Ohm's Law 14
open-loop control 8
operational amplifiers (Op Amps) 10, 18–25, 94
  characteristics 18
  circuit assessment 24–5
  uses 18
OR gate 30, 31, 37, 38, 39
output symbol 40
overloading 90

parallel circuits 15
'pause' command 42
PBASIC 42–5
  counters 44–5
  'if' command 44
  writing and understanding code 42–3
plain bearings 56
plastic range 85
plasticity 84
pneumatic systems 60–7
  automatic circuits 66
  cascade systems 66–7
  cylinders 60–1
  sequential control 64–5, 94
  speed control 62–3
  time delay 63
  valves 60–1
PNP transistor 26
positive feedback 11
potential energy 13
power 12
  in circuit 14
  mechanical 58–9
programmable control 40–51
  flowcharting 40–1, 51, 94
  motor control 50–1
  writing code 42–9
proportional control 10
Pulse-Width Modulation (PWM) 50–1, 94
push-pull driver 94

INDEX

question paper 5

rack and pinion 53
reactions 68, 76, 77
redundant member 82, 94
reference level 10
reference voltage 10
reservoir 63
restrictors 60–1
resultants 74, 94
rigid couplings 54
roller bearings 57
rubber 55, 84

safe working load (SWL) 91
safe working stress 91
saturation 27, 28–9
seat heating system 45
semi-automatic circuits 66
sequential control 64–5, 94
series circuits 15
shafts 54, 94
shuttle valve 60
significant figures 5, 19, 94
simple gear train 52
single acting cylinder 60
skills 94
   knowledge versus 7
social impacts 7
soft start 51, 94
source 28, 29
space time 51, 94
speed control 62–3
spider 55
split bearings 57
Stamp controllers 42–5
start/stop symbol 40
static equilibrium 68, 75, 80, 94
steel 84

strain 86–7
street-lighting system 9
strength 84
stress 86, 87
stress–strain graphs 87
structural engineering 6
structural failure, reasons for 90–1
structures 68–83
   framed 78–9
   hinged supports 76–7
   nodal analysis 80–3, 94
   reactions 68, 76, 77
   resultants 74, 94
   uniformly distributed load (UDL) 69
   see also forces; moments
struts 78–9, 80–1, 82, 94
sub-procedures 40, 41
sub-system diagrams 8–9
summing amplifier 22
supports
   hinge and roller 76
   loaded 76–7
syllabus 4
'symbol' command 44–5
symbols 40
syntax, importance 43, 47

tensile forces 78–9, 94
tensile test 85
thermistor graph 17
thrust bearings 57
ties 78–9, 80–1, 82, 94
time delay 63
toothed belt 53
torque 58
traffic lights 43, 47, 79
transconductance 29
transducer graphs 17

transistors 26–7
   amplification by 26
   in circuits 26–7
triangulation 78
truth tables 34–6, 94
tumble drier 39
two-state control 10

ultimate load 85, 91
ultimate tensile stress (UTS) 87, 91
unidirectional restrictor 60–1
uniformly distributed load (UDL) 69
universal joint 55
Universal System Diagram 8

V belt 53
valves 60–1
vectors, sum of 74
'void loop()' command 46–7
'void setup()' command 46–7
voltage follower 23
voltage-divider circuits 16–17, 94

wait symbol 40
wall bracket 77
weight 12
work done 12
worm and nut 52–3
worm and wheel 52

XOR gate *see* EOR gate

yield point 85
yield strain 87
yield stress 87
yield stress point 87
yoke 55
Young's Modulus 88–9